THE SECD MICROPROCESSOR

THE KLUWER INTERNATIONAL SERIES IN ENGINEERING AND COMPUTER SCIENCE

VLSI, COMPUTER ARCHITECTURE AND DIGITAL SIGNAL PROCESSING

Consulting Editor
Jonathan Allen

Latest Titles

THE SECD MICROPROCESSOR

A Verification Case Study

by

Brian T. Graham
University of Calgary
presently at the University of Cambridge

KLUWER ACADEMIC PUBLISHERS
Boston/Dordrecht/London

Distributors for North America:
Kluwer Academic Publishers
101 Philip Drive
Assinippi Park
Norwell, Massachusetts 02061 USA

Distributors for all other countries:
Kluwer Academic Publishers Group
Distribution Centre
Post Office Box 322
3300 AH Dordrecht, THE NETHERLANDS

Library of Congress Cataloging-in-Publication Data

Graham, Brian T., 1951-
    The SECD microprocessor : a verification case study / by Brian T.
Graham.
        p.   cm. -- (The Kluwer international series in engineering and
    computer science ; #178)
        Includes bibliographical references (p.   ) and index.
        ISBN 0-7923-9245-0 (acid-free paper)
        1. Microprocessors. 2. Computer architecture.   I. Title.
    II. Series: Kluwer international series in engineering and computer
    science ; SECS 178.
    QA76.5.G656   1992
    621.39'2--dc20   92-12620
                                                                      CIP

Printed on acid-free paper.

Printed in the United States of America

Contents

Foreword

This is a milestone in machine-assisted microprocessor verification. Gordon [20] and Hunt [32] led the way with their verifications of simple designs, Cohn [12, 13] followed this with the verification of parts of the VIPER microprocessor. This work illustrates how much these, and other, pioneers achieved in developing tractable models, scalable tools, and a robust methodology. A condensed review of previous research, emphasising the behavioural model underlying this style of verification is followed by a careful, and remarkably readable, account of the SECD architecture, its formalisation, and a report on the organisation and execution of the automated correctness proof in HOL.

This monograph reports on Graham's MSc project, demonstrating that — in the right hands — the tools and methodology for formal verification can (and therefore should?) now be applied by someone with little previous expertise in formal methods, to verify a non-trivial microprocessor in a limited timescale. This is not to belittle Graham's achievement; the production of this proof, working as Graham did from the previous literature, goes well beyond a typical MSc project. The achievement is that, with this exposition to hand, an engineer tackling the verification of similar microprocessor designs will have a clear view of the milestones that must be passed on the way, and of the methods to be applied to achieve them.

The reader should be warned that formal verification is not easy; I am sure Graham endured many frustrations elided in this beguilingly effortless presentation. The task appears surprisingly easy only because the problem has been broken down in a way that manages complexity. This is a model for you to follow in attacking a class of verification problems. If your problem differs significantly from the example, you may come to appreciate how easy it is to propagate, rather than contain complexity.

No book can cover all aspects of the formal verification. Anyone seriously embarking on a verification project should consider some aspects in greater depth. This is a report on the application of a particular model of behaviour to analyse a particular system using the HOL proof assistant. You will want to analyse other systems and should give careful consideration to the choice of models and tools. Any formal model of behaviour fails to correspond to the behaviour of real hardware in many subtle ways — the engineer embarking on a novel verification exercise should be aware of the limitations of the model, and of the appropriateness of the model to the analysis required. In considering other tools — both other types of tool (it is sheer masochism to use an interactive theorem prover when automated tools will apply), and other theorem provers — a myriad of different considerations impinge. This book provides little assistance in taking such decisions. The reader should take this as a warning, not a criticism. In a text of this size and scope, some things must be abridged. I cannot fault the balance of the presentation, nor the choice of model and tool for this example.

Michael P. Fourman

Preface

Great strides in the technology of fabrication of integrated circuits
have created a gap between production capability and the ability
to manage the complexity that such devices invite. As the smallest
feature size of integrated circuit devices drops, the number of transistors that can be integrated into a single design increases. Current
microprocessor designs have in excess of one million transistors on a
single chip[1]. Quite apart from simple transistor count, the devices are
themselves increasingly complex and the development of massively
parallel systems promises that this trend will continue. Reasoning
about things of such complexity demands a high degree of formalism
in order to ensure the behaviour of the device is understood. The
informal methods used for design in the past are not always adequate
for today's designs.

The decrease in cost coupled with the increasing functionality of
integrated circuit devices has led to a rapid growth in their range
of application. Today we find them in such diverse locations as the
control systems of aircraft, automobiles, ships, and trains; medical
devices such as pacemakers and artificial limbs; embedded within
"smart cards" for banking, medical records, and other applications;
industrial control systems including nuclear generating stations; remote sensing systems on pipelines; as well as assorted military systems.

The potential for a flawed design to cause economic disaster or
loss of life is very real. Numerous documented cases of risk and actual
loss connected with computer systems and related technology have
appeared in the pages of the ACM SIGSOFT quarterly, Software
Engineering Notes, spread over many issues. Peter Neumann has
produced a summary of these cases with over 500 entries [46]. Loss

---

[1]The Intel 80486 with 1.1 million and the Motorola 68040 with 1.2 million
transistors were expected to be available in mid 1990, but when the 80486 was
marketed in late 1989, production was suspended when an error was discovered
in the floating point unit despite what was described as "hundreds of thousands
of hours" of testing.

and risk categories include loss of life, potentially life-critical, loss of resources, and security/privacy/integrity problems. Behaviours to which the losses are attributed include intentional misuse; accidental misuse; misinterpretation/confusion at a man-machine interface; flaws in system concept, requirements, design, or implementation; improper maintenance/upgrade; and hardware malfunction attributable to system deficiencies, electronic or other interference, the physical environment, acts of God, etc.

The traditional method of ensuring the correctness of designs is simulation. This entails defining a model of the behaviour of the primitive components of the design and exhaustively determining that the behaviour for each possible input condition is as desired. As the number of cases is exponential in the number of input bits for combinational circuitry, the increasing device size has made total coverage simulation too time consuming to be acceptable. Devices which maintain internal state are far more difficult (to the point of impossible) to simulate exhaustively for each state and each input condition. Thus products are being sold today with less than a desirable level of assurance of behaviour over the range of all input conditions.

Formal methods, entailing the description of and reasoning about systems within a formal logic, are being applied to all levels of systems design, including both software and hardware components, in an effort:

> ...to increase the quality of the systems developed and to increase our confidence that the systems will behave in a predictable manner.[2]

The nature of this research

The goal of this research was to examine the application of formal methods to the design of complex integrated circuits. This work was part of a larger project on the use of formal methods in systems design at the University of Calgary. The design of hardware is only one part of producing reliable systems. It is equally important to

---

[2]Dan Craigen in [14].

assure the correctness of the software which will run on the system, and most important of all is the interface between the two. It is most desirable if a common formalism can be used to express both. At Calgary, the VLSI group chose to restrict its attention to functional languages running on functional architectures. It is not that the hardware is any easier to verify, but proofs of program correctness certainly are, as is the verification of the translation process. The key points in the approach are:

1. Use a functional programming language. Since they are based upon the λ-calculus they are expressive. They are also very succinct and amenable to proof.

2. Use a sugared variant of λ as the compiler target language. Since functional constructs are easy to express in terms of λ, it is relatively straightforward to express the semantics of the translation scheme and to prove its correctness.

3. Convert from λ to machine code. This step is relatively trivial if we choose a functional architecture, e.g. SECD which supports λ, or a graph reduction machine which will execute combinators.

4. Run the code on verified hardware.

Choosing λ as the common thread considerably simplifies all the above step-by-step transitions. In particular, it is possible to verify software and hardware with the same proof checker, to adopt a single proof style, and to reuse proofs.

The work is part of a long term effort in verification that started in 1985. The VLSI group chose to work with LispKit [28, 29, 30, 50], and Henderson's version of Landin's abstract SECD machine [28]. [7, 17, 28, 31] explain the workings of varieties of eager and lazy SECD machines. [17] sketches Plotkin's [47] proof of correctness of an eager SECD machine. Thus the choice of SECD was deliberate — there was much work to build on. To date the project group has

- designed, fabricated and is presently testing version II of the chip,

- constructed a rig and associated software, including compilers from LispKit to SECD so that we can download LispKit programs and run them on SECD,

- completed a (hand) proof that the abstract SECD machine executes LispKit programs correctly [51], and

- completed a machine assisted proof in HOL of the (partial) correctness of the SECD design.

The formalism chosen for this study is a higher-order logic, which has been implemented in the HOL proof assistant [22] by Mike Gordon of Cambridge University. The use of a higher-order logic permits specifications to be succinct and often elegant, making it easier to assure through visual inspection that the specification captures the intention.

This work has several unique properties. The scope of the project, encompassing both circuit design and specification/verification in this project has led to a close resemblance between the formal specifications and the informal models that are used in the design process. The choice of a functional architecture subject provided increased complexity in defining the effect of machine instruction execution, and required the representation of abstract S-expression data structures, so that the specification operates well above the level of *bits*.

One of the most significant aspects of the project has been the size and complexity of the subject system. *Because of the complexity of the SECD chip, it is not possible within the scope of a thesis to give more than an outline of most of its component specifications and the proofs: they are simply too large to be included in their entirety. All we can do is give a flavour of the work.* Despite this incompleteness, the critical concepts in designing the specification are presented in detail, and the description of the proof strategy is augmented with representative samples of results, and quite detailed descriptions of the methodology. A more complete examination of the specifications and proofs requires the HOL sources, and these have been made freely available. Information on obtaining them is given on page 58. We hope that this will provide a useful source of nontrivial teaching examples and problems. The impact of the project size, particularly on proof management, is a recurring theme. The huge size of the proof meant that many original proof management techniques had to be developed. The achievement of the proof alone stands as a significant result. The SECD chip is one of the largest examples to date in the field of hardware verification.

The structure of the book

Chapter 1 describes what is entailed in the use of formal methods in hardware design, and looks briefly at other significant works in the field. It concludes with a brief description of the HOL system, which is used for the formal definition and verification in this study. In Chapter 2, the SECD architecture is described, showing how it can support the execution of a Lisp-like high level language. Chapter 3 describes the evolution of the SECD design to the physical layout stage, detailing the development of both the external and internal architecture. A formal specification for the SECD system is defined in Chapter 4, using introduced data types to represent S-expressions, with a suite of operations on the data type objects. A definition of the implementation at a register transfer level is also defined, following the layout hierarchy as closely as possible. The chapter closes with the definition of abstraction functions which relate the two levels of description. The proof of correctness relating two levels is described in Chapter 5. The proof is undertaken under quite specific constraints, on both the interaction with the environment and the internal state of the machine. The proof is undertaken in five stages, beginning with unfolding the component definitions, followed by the evaluation of the system behaviour at each microcode instruction, corresponding to a a single clock cycle, and then onto the effect of computation of microcode sequences. Under the defined operating constraints, a liveness property of the system is proved, and finally, the last stage compares the specification and implementation behaviours. The final chapter comprises observations and conclusions drawn from the work.

The research contained in this book was originally presented in two items, the author's MSc Thesis at the University of Calgary [27] and a previously unpublished project report [25]. Several of the diagrams used in this book first appeared in papers in *Hardware Specification, Verification, and Synthesis: Mathematical Aspects* [39], edited by M. Leeser and G. Brown and published by Springer-Verlag in 1989, and in *Formal Methods for VLSI Design* [53], edited by J. Staunstrup and published by North Holland in 1990. Additionally, several passages from the latter work are reproduced with added detail in this book. I have been encouraged to publish this work by several people who believe it is important to make some larger examples widely available in this field. Errors that occurred during the project are not hidden, but rather pointed out and examined for the lessons they may provide.

Acknowledgements

This work was mostly undertaken at the University of Calgary, with the completion of the proof at the Computer Laboratory at the University of Cambridge. Graham Birtwistle has been instrumental in providing encouragement, guidance and support throughout the project. My colleagues in both Calgary and Cambridge have contributed both a stimulating atmosphere for research, and a lively and pleasant environment. Jeff Joyce played a major role in the early project stages, defining the basic architecture of the chip, and developing the microcode. Wallace Kroeker was the project leader in the design of the first version of the SECD chip. Simon Williams was responsible for the majority of the layout design, and further work on the chip controller and testing of the fabricated chips. Tom Melham and Inder Dhingra provided much needed assistance in learning the HOL system. Ian Mason's work on the Semantics of Lisp provided key ideas for the formal SECD specification. This work could not have been completed without the support of the Natural Sciences and Engineering Research Council of Canada, the Alberta Microelectronics Centre, the Canadian Microelectronics Corporation, the Communications Research Establishment, and the Association of Commonwealth Universities.

Lastly, I wish to thank my wife Jean, without whose support this could not have been completed.

THE SECD MICROPROCESSOR

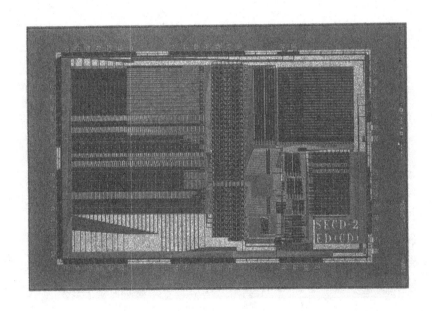

The SECD Chip: Microphotograph of Second Version

Chapter 1

Formal Methods and Verification

Formal methods involve representing an implementation and a specification within a formal theory or calculus. *Verification* compares the representations within the formal theory, reasoning that under particular constraints the implementation ensures the specification is (or is not) met.

For application to hardware, a circuit is represented by a composition of primitive components, such as simple logic gates, with levels of electrical potential abstracted to a limited number of discrete values. A primitive logic gate is expressed as a relation between its inputs and outputs. In the case of memory elements, the relation includes its state as well. Specifications are often defined in terms of more abstract data objects, in order to describe the complex behaviour in a way that may be examined for agreement with what we understand is desired. Verification involves proving that, under stated constraints representing assumptions about the operating conditions of the circuit for example, the implementation guarantees the behaviour described by the specification over all inputs and states. By using a mechanised proof system, we gain a higher degree of confidence that the proofs are indeed valid.

The use of verification must be understood in terms of both its practical and theoretical limitations. One obvious limitation is the accuracy of the model chosen for primitive components. For example, one could imagine modelling transistors as simple switches, where the transistor is either on or off, depending on the gate input. This model is appropriate at some level for CMOS, where signals are guaranteed to be strong, but will not accurately capture the behaviour of pull-up transistors or transistors with weak gate signals, and could lead to an incorrect conclusion. On the other extreme, a transistor model could be as complex as the one used in SPICE. This model can give us a much higher degree of accuracy, but complexity of the proof becomes simply unmanageable for anything other than a trivial circuit. Clearly, one must choose a model appropriate for the subject, and make explicit the assumptions under which the model can capture the subject behaviour. The exceedingly complex behaviour of transistors suggests that a higher level view of the design would be more suitable for modelling. The choice of primitive logic gates and latches as the lowest level components used to model fully complementary CMOS circuits combines simplicity with accurate capture of behaviour, given the assumption that all signals are allowed enough time to settle to stable values. Full simulation of the primitive components using such tools as SPICE can establish the detailed operating constraints under which the model will be valid.

Aside from representing primitive component behaviours, the circuit connections must themselves be correctly captured in the model. There is a need to integrate the formal model with design tools to ensure that this correspondence is maintained.

Production of a correct design requires both that the design itself meets the specification of the desired behaviour, and that the design is actually produced in silicon. The former is the realm of verification, while the latter is in the realm of testing of the product. While the two realms can be viewed as distinct, the full formal specification of the design's behaviour in conjunction with operating constraints can assist in determining an appropriate test suite for the product. Formal methods could also be used to reason about the behaviour of an assembly of exhaustively tested subcomponents.

There is necessarily a gap between the lowest level of representation and the physical hardware device, just as geometry only describes abstractions of physical objects. At the other end of the spectrum, the specification is a representation of some designer's intention, which cannot be entirely captured within a formal logic. Nor can we ever prove the validity of the specification as a representation of these intentions. These gaps are not unique to formal methods, but their presence defines the limits of what formal methods can contribute to assurance of design correctness. Thus the term "partial" should be assumed whenever the words "verification" or "correctness" appear in this work.

1.1 Achievements in Hardware Verification

The first significant achievement in hardware verification was Gordon's machine assisted proof of the correctness of a small microprocessor with a microcoded control unit [20]. This 8 instruction machine was specified at the register transfer level, and the correctness of this model meeting a higher level specification was proved. This work was done in the LCF-LSM system [19], a predecessor of the HOL system.

Warren Hunt specified and proved the correctness of the FM8501 microprocessor [32], a traditional von Neumann architecture comparable to a PDP-11 in complexity. The specification and verification was done in the first-order Boyer-Moore logic, and proved using the associated automated theorem prover.

Jeff Joyce designed the TAMARACK microprocessor based on Gordon's original example, specified and verified it in HOL [34]. He has extended this work to the transistor level [36], added a configurable memory timing interface and parameterised the specification data types and operations [35], and has since verified the correctness of a TINY compiler generating code for TAMARACK [37]. TINY (see chapter 3 in [18]) is a toy imperative language which includes assignments, conditionals, and while statements.

Perhaps the largest single verification effort to date has been the VIPER microprocessor by RSRE [15] and Cohn [12, 13]. This work is distinguished by the fact that formal methods were applied to a commercially available product, and the considerably larger size and complexity of the device and verification effort compared to the previous examples. The chip is hard wired rather than microcode controlled, and was defined at roughly a register transfer level with detailed implementation of data operations, at a major state level which described the operation of the chip in terms of a graph traversal, and at a more abstract top level. The correspondence of the two upper levels was fully verified, but the extension to the lowest of the three levels is incomplete, although a significant analysis of the implementation through proof techniques was accomplished.

Other significant efforts include the flooding sink local area network broadcast message eliminator by Melham [4], the Sobel image processing chip by Narendran and Stillman [45], and the Cayuga microprocessor by Sekar and Srivas [49]. Dhingra [16] has formalised and validated CLIC, an integrated circuit design style, in HOL.

Significant work at Computational Logic followed from Warren Hunt's work on the FM8501 described above. Bevier [2] has gone on to implement and verify a multi-tasking operating system kernel for a 16–bit von Neumann architecture which includes process scheduling, response to error conditions, message passing primitives, and character I/O. Moore [44] has specified the PITON language and mechanically verified its implementation on the FM8502 architecture via a compiler, assembler and linker. PITON is an assembly language designed for verified applications and includes recursive subroutine support, stack based parameter passing, and several abstract data types. Finally Young [55] has mechanically verified a code generator for Gypsy 2.05 down to PITON.

1.2   The HOL System

The formalisation and verification of the SECD microprocessor uses the HOL proof assistant. The choice of the HOL system and the use of a higher-order logic offers an expressive power that other formalisms lack. Clearly this is at some cost, as proof automation is an even more difficult problem than for a first-order logic such as the one used by Boyer and Moore [5]. The expressive power is essential to capture a wide range of views of a device, and to be able to relate them in understandable ways. Without this power of expression, both the specifications and the meaning of the correctness proofs may become too obscure to be useful.

The HOL system is a very widely used proof assistant in hardware verification. Not only has it been proved a reliable tool, but there is a growing body of worked examples and a large and growing library of supporting work, including extensive sets of theorems characterising defined data types such as integers, sets and bit strings.

1.2.1 A brief introduction to HOL

This chapter closes with a brief look at the HOL system used for the formal representation and verification of the SECD system. The following owes much of its organization to the example of Cohn [13]. A full description is beyond the scope of this work, and the reader is referred to the HOL manuals [8, 9, 10] for full documentation. The HOL system is not user-friendly, and requires a considerable effort to achieve proficiency in its use. The gentle introduction for the novice provided in [3] is highly recommended.

HOL is a machine implementation of a conventional higher-order logic in which problems can be expressed, and interfaced to the programming language ML in which proof procedures and strategies can be encoded. The type discipline of ML ensures that the only way to creating objects of type *thm* is by the application of inference rules to other theorems or axioms. Theorems are identified by the turnstile symbol ⊢, with assumptions to the left, and conclusion to the right.

New types, constants and axioms can be introduced by the user, and are organized in logical *theories*. Proved theorems may be saved in and retrieved from the theories, which are organized into hierarchies in which types, constants, axioms, and theorems are inherited from ancestor theories.

The HOL system uses the ASCII characters ~, /\, \/, ==>, !, ?, @, and \ to represent the logical symbols ¬, ∧, ∨, ⊃, ∀, ∃, ϵ and λ respectively. Throughout this thesis, the symbols ~, \/, /\ and \ will be replaced by the conventional logical symbols. A *term* of higher-order logic can be one of the following:

- A **variable**;

- A **constant**, including natural numbers, the boolean values T and F, etc;

- A **function application** of the form t1 t2;

- An **abstraction** of the form λx.t;

- A **negation** of the form ¬t;

- A **conjunction** of the form t1 ∧ t2;

- A **disjunction** of the form `t1 ∨ t2`;

- An **equality** of the form `t1 = t2`;

- An **implication** of the form `t1 ==> t2`;

- A **universal quantification** of the form `!x.t`;

- An **existential quantification** of the form `?x.t`;

- An **ε-term** of the form `@x.t`,[1] expressing some arbitrary value `x` such that the predicate `t` is true;

- A **conditional** of the form `t=>t1|t2`, expressing *if* `t` *then* `t1` *else* `t2`;

- A **local declaration** of the form `let x = t1 in t2`;

- A **list** of the form `[t1; t2; t3;...;tn]` where all elements have the same *type*;

- A **pair** of the form `(t1,t2)`, where `t1` and `t2` may each be of any *type*.

Double quotes distinguish HOL terms in the ML interface, and HOL terms will be identified by the consistent use of `typewriter` `font`. The ML antiquotation operator `^` permits ML identifiers bound to HOL terms to be included within HOL terms. ML comments are enclosed within `%` characters.

All terms in HOL have a *type*. The expression `t:ty` means `t` has type `ty`. Built-in types include `:bool` and `:num` for *booleans* and *natural numbers*. Three type operators are `->`, `+`, and `#`, for describing function types, sum types, and product types respectively. Types may be parameterised, for example `:(bool)list` is the type of *boolean* lists. Polymorphism is allowed, and type variables are typically `*`, `**`, and so on. Nonempty new types may also be defined by mapping to an existing type within the logic. Types will consistently be shown preceded by a colon.

---

[1] `@` is a higher-order version of Hilbert's choice operator, called the "SELECT" operator in HOL.

Many constants are built into the HOL system, including the boolean constants and natural numbers, arithmetic operators +, -, *, <, <=, >, >=, DIV, MOD, EXP, SUC, and PRE, the list operations CONS, HD and TL, and FST and SND selectors on pairs, to name some of the more commonly used ones. The reader's attention is particularly directed to the infix function composition operator o, which is used repeatedly in abstraction functions.

There are two general approaches to proof within HOL: *forward* and *backward*. *Forward* proof works by applying inference rules to existing theorems and axioms to derive a desired result in the form of a new theorem. Where to start on a large complex proof is problematic, and managing the many branches involved is oftentimes exceedingly difficult. An alternative methodology, *backward* proof, starts with a statement of the theorem you would like proved (a *goal*), which the user incrementally splits into smaller, more manageable subgoals. The HOL system manages the state of the proof on a goal stack, and when each subgoal is reduced to a theorem, it assembles the entire proof and returns the desired theorem. This methodology does not provide a distinct means of construction of (ML) *thm* type objects, rather it allows the user to generate the proof, which will use the same inference rules as the *forward* proof methodology with a top-down approach, leaving the system to manage the details. Both methods have advantages and both are used in this work. The choice of *forward* or *backward* proof will often be an important consideration in the methodology.

The HOL notation is used for the formal definition of the SECD system, but the preceding description should make the material accessible to users familiar with other machine proof systems. Before proceeding with a formal definition, we present an informal description of the abstract SECD architecture, followed by a description of the development of the SECD chip design.

Chapter 2

LispKit and the SECD Architecture

In this chapter, we shall describe somewhat informally the abstract SECD architecture which the microprocessor design is intended to implement. To illustrate the suitability of the architecture for supporting execution of functional language programs, a high-level programming language, LispKit, is introduced, and a schema for translating LispKit programs to machine code is given. The translation schema will be helpful in understanding the architecture, in particular the instructions for function declarations and calls. The definitions of LispKit and SECD are taken from Henderson [28]. Henderson also defines a LispKit interpreter, and a LispKit to SECD compiler.

LispKit is a pure functional subset of the Lisp language. A pure functional subset means that LispKit has no destructive assignment operation, so that the value of an expression is uniquely determined by the value of its constituent parts, and identical expressions always have the same value. S-expressions are defined, and then a syntax for LispKit is given. This is followed by an informal semantics for the language, defined by an interpreting function written in Franz Lisp. Issues fundamental to supporting the language with a hardware system such as bindings and representation of function-valued objects are discussed.

The SECD machine architecture is described by its machine instructions and state transitions effected by each, giving a semantics for the machine language. Following this, a translation schema for well-formed LispKit expressions into SECD machine code is the basis for a LispKit compiler, and defines an operational semantics for LispKit.

Different fonts will be used to help distinguish the different languages presented: *Roman Italics* for Franz Lisp expressions, *Sans Serif Italics* for LispKit expressions, and Sans Serif for SECD machine code expressions.

2.1   The Syntax of LispKit

Fundamental to the Lisp programming world is the class of objects known as symbolic expressions, or S-expressions for short. These are defined recursively as:

S-expression ::= atom | (S-expression . S-expression)

Atoms are of two types: numeric and symbolic. A numeric atom is a possibly signed sequence of digits, which is taken as representing a decimal integer. Symbolic atoms, either constants or variables, appear as a series of letters or digits or other characters, beginning with a character. There are three special symbolic atoms: *NIL*, *T*, and *F*, which are symbolic constants and have a particular meaning attached to them.

A dotted pair is the result of a Lisp *cons* operation on two S-expressions. If *a* and *b* are S-expressions, then *(cons a b)* produces a dotted pair *(a . b)*. The dot notation used here is often replaced by the list form where possible. The rules for transforming from dot to list notation are simply:

- *(a . NIL)* may be written as *(a)*

- *(a . (b))* may be written as *(a b)*

where *a* and *b* may be any S-expressions. *NIL* is interpreted as the empty list.

LispKit provides sixteen primitive operators which are reserved symbolic constants. In order to distinguish constants from variables, constants other than the sixteen operators are represented by the dotted pair whose first component is the *QUOTE* operator. Structural operators are *CONS, CAR, CDR*, and *ATOM*, which perform the standard list operations common to all Lisp variants. Arithmetic operators include *ADD, SUB, MUL, DIV* and *REM*. The relational operators are limited to *EQ* and *LEQ*. In addition, there is the conditional operator *IF*, the λ operator *LAMBDA*, used for defining functions, and two block defining operators, *LET* and *LETREC*, the latter being used for recursive bindings.

Well-formed expressions in LispKit form a subset of the set of S-expressions, as determined by the derivation rules of Table 2.1.

x	variable
(QUOTE s)	constant
(ADD e_1 e_2)	
(SUB e_1 e_2)	
(MUL e_1 e_2)	arithmetic expressions
(DIV e_1 e_2)	
(REM e_1 e_2)	
(EQ e_1 e_2)	relational expressions
(LEQ e_1 e_2)	
(CAR e)	
(CDR e)	
(CONS e_1 e_2)	structural expressions
(ATOM e)	
(IF e_1 e_2 e_3)	conditional form
(LAMBDA (x_1 ...x_n) e)	λ-expression
(f e_1 ...e_k)	function call
(LET e (x_1.e_1) ...(x_k.e_k))	simple block
(LETREC e (x_1.e_1) ...(x_k.e_k))	recursive block

where e, e_i are well-formed expressions,
   x, x_i are symbolic atoms (variables),
   s is any S-expression,
   f is a λ-expression.

Table 2.1: Well-Formed LispKit Expressions

- **Constants** may be other than just numbers, as the restriction to S-expressions implies. However, all constants must be preceded by the *QUOTE* operator. The **arithmetic and relational operators** are all binary operators, unlike those in many variants of Lisp. The **structural expressions** are typical of most Lisps.

- The **conditional form** takes three arguments, the first of which is the conditional expression. The second argument is evaluated only if the conditional expression evaluates to *T*, otherwise the third is evaluated.

- **Lambda expressions** are used for defining functions of one or more arguments. A list of λ-bound variables is followed by an expression which will usually include all occurrences of these variables.

- **Function applications** follow the form of primitive operators, with a function-valued expression being the first item in a list, followed by the values to be bound to its local variables.

- **Blocks** are another means of creating local bindings. The expression containing the bound variables follows immediately after the *LET* operator, followed by any number of dotted pairs of variables and expressions to be bound to the variables. Note that this syntax differs from that used by many other versions of Lisp. The only difference between the *LET* and the *LETREC* is that the expressions to be bound to the variables in the *LETREC* may include (recursive) references to any of the locally bound variables, while in a *LET* they do not reference any of the locally bound variables. We shall see, in fact, that the *LET* form is unnecessary, as it is equivalent to the function application of a λ-expression. The *LETREC*, however, is needed to define recursive functions. Furthermore, each expression e_i bound to a variable within a *LETREC* must evaluate to a function-valued object. The meaning of a function-valued object will become clear when the interpretation of LispKit is discussed. This restriction eliminates expressions to which we do not wish go give a meaning such as: *(LETREC x (x ADD (QUOTE 1) x))* which attempts to define *x* as its own successor.

Despite the derivation rules above, those expressions which will give rise to meaningful computations have not as yet been precisely defined.[1] For the most part, the primitive operators, as well as any functions we may define, are partial. For example the expression *(ADD (QUOTE 2) (QUOTE (A B C)))* is not meaningful in LispKit, because *ADD* requires that its arguments be integer-valued. Similarly, *(CAR (QUOTE NIL))* is not meaningful, since the *CAR* operation is only defined on dotted pairs.

2.2   The Interpretation of LispKit

A full interpreter for the LispKit language adapted from Henderson's [28] to Franz Lisp is given in Table 2.2. A brief description of its more interesting features follows. Two sample programs give a taste of the language in use.

Bindings of variables within LispKit are represented using the concept of contexts. A context consists of a set of bindings that associate variables with values, implemented using isomorphic lists of variable names and values. The value of a variable is the value located in the position in the *valuelist* that corresponds to the location of the variable in the *namelist*, with the added restriction that if the variable occurs more than once, the location closest to the front of the list is used. This permits new bindings to override existing bindings within a context. For example, the namelist and valuelist:

namelist: *((x y) (z x))*
valuelist: *((1 3) (5 NIL))*

represent the bindings: $x \leftrightarrow 1, y \leftrightarrow 3, z \leftrightarrow 5$. The second occurrence of *x* is rendered inaccessible in this example. This environment could have been generated by a LispKit program of the form:

```
(LET (LET (...)
        (x.(QUOTE 1)) (y.(QUOTE 3)))
    (z.(QUOTE 5)) (x.(QUOTE NIL)))
```

LispKit expressions will always be evaluated within some context. New bindings are added to the existing context by *LET* and

---

[1]The interested reader can find a complete denotational semantics of LispKit by Simpson in [51].

LETREC operators. Thus, the interpretation of a variable x in a context (n, v), is simply the value in the location in v that corresponds to the location of x in n. The interpretation of a *LET* expression is the interpretation of the expression part in the current context extended by adding the list of bound variables to the front of the variable namelist, and adding the list of values, obtained by evaluating the value expressions in the existing context, to the front of the valuelist. Thus

$$(EVAL \ '(LET \ e \ (x_1.e_1)...(x_k.e_k)) \ n \ v) =$$
$$(EVAL \ e \ ((x_1...x_k).n) \ (((EVAL \ e_1 \ n \ v)...(EVAL \ e_k \ n \ v)).v).$$

Function definitions may contain both free and bound variables within the body of a λ-expression. Bindings are defined to be static, so that values bound to free variables within a λ-expression are determined from the context in which they are defined, rather than in the context in which the function is called. To facilitate this, the notion of a closure is introduced for the interpretation of functions. The closure will consist of the defining context, along with the list of λ-bound variables, and the body of the λ-expression.

$$(EVAL \ (LAMBDA \ (x_1...x_n) \ e) \ n \ v) = (((x_1...x_n).e).(n.v))$$

The defining context consists of the namelist and valuelist. The namelist is the collection of bound variable names, and the valuelist is the corresponding collection of the values associated with each variable name. When the expression is applied to a list of arguments, the list of λ-bound variables will be added to the namelist. The evaluated arguments are added to the valuelist, creating a new context for evaluating the body of the function, within which the local variables are bound.

In a *LETREC* expression, the variables are to be bound to (possibly mutually) recursive functions, thus the context in which these functions are evaluated must include the values of the functions themselves. This is accomplished by evaluating the expressions in a context which has the values of the recursively bound variables still pending. In practice, an empty list is used as a place-holder for the pending values at the beginning of the valuelist. Since each expression is required to evaluate to a function-valued object, *i.e.* a closure, with the context parts of all being identical. Thus, a single destructive *rplaca* operation[2] can be used to alter the pending value of the valuelist to instead point to the list of closures that are created. A circular data structure is thus created. A simplified example illustrates the idea.

(EVAL (LETREC e $(f_1.e_1)\ldots(f_k.e_k))$ n v) =
(EVAL e (y.n) (rplaca v' z)

where y = $(f_1\ldots f_k)$
where v' = (PENDING.v)
where z = $((EVAL\ e_1\ (y.n)\ v')\ldots(EVAL\ e_k\ (y.n)\ v'))$

The closures formed by evaluating $e_1\ldots e_k$ each contain the valuelist with the *PENDING* placeholder. The *rplaca* operation not only builds the valuelist for the context in which to evaluate the expression *e*, but also installs the closures in the valuelist of each closure as well. The reader should recognize that this use of a destructive Franz Lisp operation in defining the interpretation of the LispKit expression does not conflict with the status of LispKit as a purely functional language. The LispKit programmer does not have a destructive operator available for use in programming.

To complete the interpretation, we consider each remaining well-formed LispKit expression.

- **Constants** are represented by a dotted pair with the atom *QUOTE* as the *car*. Regardless of its context, it will evaluate to the *cadr* of the pair. Thus *(QUOTE 2)* evaluates to *2*, and *(QUOTE (a b c))* evaluates to *(a b c)*. The three special symbolic atoms, *NIL*, *T*, and *F* are mapped to the Franz Lisp values of *nil*, *t*, and *nil* respectively in the interpreter.

- The **arithmetic operators** work as one would expect. For example, *(ADD e_1 e_2)* in the context *(n,v)* will evaluate to the sum of the values of e_1 and e_2, both evaluated in *(n,v)*. *SUB, MUL, DIV,* and *REM* work similarly.

---

[2]The *rplaca* function replaces the *car* of its first argument with its second argument by overwriting the top *cons* cell of the first argument.

```
(def EVAL
    (lambda (e n v)
      (if (atom e)                                              ; variable
          (assoc e n v)
        (let ((key (car e)))
          (if (eq key (quote QUOTE))                            ; constant
              (let ((const (cadr e)))
                (if (eq const 'NIL)    nil
                (if (eq const 'TRUE)  t
                (if (eq const 'FALSE) nil const)))
          (if (eq key (quote ADD))
              (+ (EVAL (cadr e) n v) (EVAL (caddr e) n v))
          (if (eq key (quote SUB))
              (- (EVAL (cadr e) n v) (EVAL (caddr e) n v))
          (if (eq key (quote MUL))
              (* (EVAL (cadr e) n v) (EVAL (caddr e) n v))
          (if (eq key (quote DIV))
              (/ (EVAL (cadr e) n v) (EVAL (caddr e) n v))
          (if (eq key (quote REM))
              (rem (EVAL (cadr e) n v) (EVAL (caddr e) n v))
          (if (eq key (quote EQ))
              (eq (EVAL (cadr e) n v) (EVAL (caddr e) n v))
          (if (eq key (quote LEQ))
              (<= (EVAL (cadr e) n v) (EVAL (caddr e) n v))
          (if (eq key (quote CAR)) (car (EVAL (cadr e) n v))
          (if (eq key (quote CDR)) (cdr (EVAL (cadr e) n v))
          (if (eq key (quote CONS))
              (cons (EVAL (cadr e) n v) (EVAL (caddr e) n v))
          (if (eq key (quote ATOM)) (atom (EVAL (cadr e) n v))
          (if (eq key (quote IF))
              (let ((e1 (cadr e)) (e2 (caddr e)) (e3 (cadddr e)))
                (EVAL (if (EVAL e1 n v) e2 e3) n v))
          (if (eq key (quote LAMBDA))
              (cons (cons (cadr e) (caddr e)) (cons n v))
          (if (eq key (quote LET))
              (let ((y (vars (cddr e))) (z (evlis (exprs (cddr e)) n v)))
                (EVAL (cadr e) (cons y n) (cons z v)))
          (if (eq key (quote LETREC))
              (let* ((y (vars (cddr e)))
                     (v1 (cons (quote PENDING) v))
                     (z (evlis (exprs (cddr e)) (cons y n) v1)))
                (EVAL (cadr e) (cons y n) (rplaca v1 z)))
            (let ((c (EVAL key n v)) (z (evlis (cdr e) n v)))      ; application
              (EVAL (cdar c) (cons (caar c) (cadr c)) (cons z (cddr c)))
          )))))))))))))))))))))

(def APPLY
    (lambda (f x)
      (let ((c (EVAL f nil nil)))
        (EVAL (cdar c) (cons (caar c) (cadr c)) (cons x (cddr c))))))))
```

Table 2.2: LispKit Interpreter Written in Franz Lisp

- The **relational operators**, *EQ* and *LEQ*, evaluate their arguments in the same fashion as the arithmetic operators. The simplified interpreter treats *EQ* as effecting the same operation as the *eq* function in Franz Lisp, in that two S-expressions are equal if they are both atoms and they evaluate to the same value, or if they are both dotted pairs and they are both pointers to the same cons cell. This differs from the definition given by Henderson ([28] pp. 22, 53), which defines *EQ* only when at least one of its arguments is an atom. The original meaning will be kept in subsequent sections.

- The **structural operators**, *CAR, CDR, CONS* and *ATOM* are interpreted as performing the same operations as the corresponding *car, cdr, cons* and *atom* operations in Franz Lisp, when applied to the interpreted arguments. Again, arguments are evaluated as for arithmetic operators.

- The **conditional form** evaluates its first argument in the given context, and if this evaluates to *T*, then the second argument is evaluated; otherwise the last argument is evaluated. This form evaluates only one of the two branches, permitting, for example, testing for terminating conditions of recursive function definitions.

- The **function application** is interpreted by adding the value of the function's arguments, interpreted in the current context, to the start of the valuelist in the context part of the closure. Similarly, the list of bound variables is added to the start of the namelist in the context part of the closure, and the body of the closure is evaluated in the thus extended context.

Finally, the top level program must evaluate to a function-valued object which is applied to a list of arguments. Free variables are not permitted in the top level program, so that our starting context consists of a pair of empty lists.

Example 2.1 *(twice double)* function

The first example is a nonrecursive curried function that takes a function as an argument and returns a function-valued object:

```
(LET (twice double)
     (double LAMBDA (x) (ADD x x))
     (twice LAMBDA (f) (LAMBDA (x) (f (f x)))))
```

which shall be applied to the argument list *(7)*. Let *example-1* represent this expression, and e_1 and e_2 the expressions to be bound to *double* and *twice* respectively.

The evaluation of *(APPLY example-1 '(7))* begins by evaluating *example-1* in the empty context. The first call to *EVAL* has the keyword *LET* as the first atom, and thus the list of bound variable names *(twice double)*, and the list of their evaluations is used to create a context in which the expression *(twice double)* is evaluated.

$$(EVAL\ example\text{-}1\ ()\ ())$$
$$\Longrightarrow\ (EVAL\ (twice\ double)$$
$$((double\ twice))$$
$$(list\ (cons\ (EVAL\ e_1\ nil\ nil)$$
$$(cons\ (EVAL\ e_2\ nil\ nil)\ nil))))$$

The two expressions, e_1 and e_2, evaluate to function closures as follows:

$$(EVAL\ e_1\ nil\ nil)$$
$$\Longrightarrow\ (((x)\ ADD\ x\ x)\ nil)$$
$$(EVAL\ e_2\ nil\ nil)$$
$$\Longrightarrow\ (((f)\ LAMBDA\ (x)\ (f\ (f\ x)))\ nil)$$

The next stage evaluates the application *(twice double)* in the context extended with the new variables and closure values by evaluating the function part and argument parts separately. Both *twice* and *double* are variables, so their values are simply fetched from the context.

$$\Longrightarrow\ (EVAL\ (twice\ double)$$
$$((double\ twice))$$
$$((\ (((x)\ ADD\ x\ x)\ nil)$$
$$(((f)\ LAMBDA\ (x)\ (f\ (f\ x)))\ nil)\ )))$$

The body of the function valued object *(LAMBDA (x) (f (f x)))* (*i.e. twice*) is evaluated in a new context extended with the variable *f* and the value of *double*, producing a closure as the final result of evaluating *example-1*.

$$\Longrightarrow \; (EVAL \; (LAMBDA \; (x) \; (f \; (f \; x)))$$
$$((f))$$
$$(((((x) \; ADD \; x \; x) \; nil))))$$
$$\Longrightarrow \; (((x) \; f \; (f \; x))$$
$$((f))$$
$$((((x) \; ADD \; x \; x) \; nil)))$$

The closure parts include the list of bound variables *(x)*, the body of the function *(f(f x))*, the namelist *((f))* and the valuelist *((((x) ADD x x) nil))*.

Returning to the evaluation of the top level *APPLY* function, the body of the function is evaluated in a context with *(x)* and *(f)* in the variables list, and the values *(7)* and *((((x) ADD x x) nil)))* in the values list.

$$\Longrightarrow \; (EVAL \; (f \; (f \; x))$$
$$((x) \; (f))$$
$$((7) \; (((((x) \; ADD \; x \; x) \; nil)))))$$

The variable *f* is bound to a function closure, but its argument *(f x)* must be evaluated. This is done in the same context, so that *f* is bound to the function closure of *double* and *x* to 7.

$$(EVAL \; (f \; x)$$
$$((x) \; (f))$$
$$((7) \; ((((x) \; ADD \; x \; x) \; nil)))))$$
$$\Longrightarrow \; (EVAL \; (ADD \; x \; x) \; ((x))((7)))$$
$$\Longrightarrow \; (+ \; (EVAL \; x \; ((x))((7)))$$
$$(EVAL \; x \; ((x))((7))))$$
$$\Longrightarrow \; (+ \; 7 \; 7)$$
$$\Longrightarrow \; 14$$

The result of this evaluation is installed in the context as the value to be bound to *x* when the outer function application is evaluated.

$(EVAL (f (f x))$
$\qquad ((x) (f))$
$\qquad ((7) ((((x) ADD x x) nil))))$
$\Longrightarrow (EVAL (ADD x x) ((x)) ((14)))$
$\Longrightarrow (+ (EVAL x ((x)) ((14)))$
$\qquad (EVAL x ((x)) ((14))))$
$\Longrightarrow (+ 14 14)$
$\Longrightarrow 28$

Example 2.2 *even* function

The next example is the *even* function, defined using mutual recursion.

```
(LETREC even
            (even LAMBDA (x)
                    (IF (EQ x (QUOTE 0))
                        (QUOTE TRUE)
                        (odd (SUB x (QUOTE 1))))))
            (odd LAMBDA (x)
                    (IF (EQ x (QUOTE 0))
                        (QUOTE FALSE)
                        (even (SUB x (QUOTE 1)))))))))
```

To apply the function to the argument list *(1)*, first evaluate the *LETREC* expression. Both of the items bound therein are evaluated in a context with the namelist augmented by *(even odd)* and the initially empty valuelist onto the start of which is *cons*'ed the special *PENDING* atom. Since both expressions are λ-expressions, they will evaluate to function closures, both containing identical contexts (in fact, pointers to the same context), with *PENDING* as the first item in the valuelists.

The body of the *LETREC*, the bound variable *even*, is then evaluated in the same context, but modified by destructively replacing the *PENDING* atom with the list of closures obtained for the λ-expressions above. This will create a circular valuelist structure, so a graphical representation will be used.

```
(EVAL example-2 nil nil)
⟹ (EVAL even
          ((even odd))
          (rplaca (PENDING)
                  ((((x)
                      IF (EQ x (QUOTE 0))
                         (QUOTE TRUE)
                         (odd (SUB x (QUOTE 1))))
                    ((even odd))
                    PENDING)
                  (((x)
                      IF (EQ x (QUOTE 0))
                         (QUOTE TRUE)
                         (even (SUB x (QUOTE 1))))
                    ((even odd))
                    PENDING)))
```

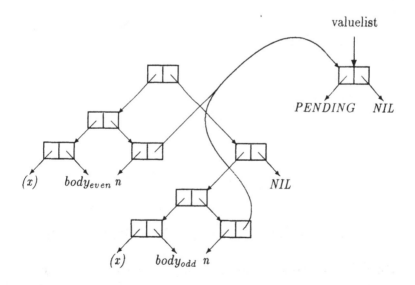

Figure 2.1: Valuelist Structure Before Destructive Operation

The important thing to note in Figure 2.1 is that the atom *PEND-*

ING occurs in only one place, with pointers to it from several places. Thus, doing a destructive *rplaca* operation on the list with *PEND-ING* as the first item, will change the value of *PENDING* in **all** locations, and create a circular list structure, so that references to the recursive functions within the body of the functions, properly refer to the closure value used to represent the function.

The next step consists of evaluating the expression *even*, which returns a closure from the valuelist. Evaluating the application of this function to the argument list *(1)*, the body of *even* is evaluated in a context augmented with the binding *1* to the λ-bound variable *x*. It should be readily apparent from the valuelist of Figure 2.2 that recursive references to *even* and *odd* are possible as the appropriate function closures are included.

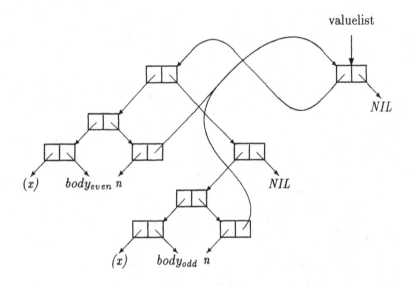

Figure 2.2: Circular Valuelist Structure After Destructive Operation

Once the circular environment structure has been established, the remainder of the function evaluation continues in a similar fashion to the previous nonrecursive example.

These examples complete the description of the LispKit language. In summary, the language has been specified by giving a syntax, and then by means of an interpreting function a semantics as well. The concept of closures to represent the value of a function object, and the use of context to define bindings, was illustrated by the two examples, using both higher order and recursive functions.

2.3   SECD Architecture

Next we give a detailed description of the SECD machine architecture, and show that it will support the execution of programs compiled from LispKit Lisp. A complete description of the machine instructions and state transitions effected by each will define the semantics of the machine language.

The SECD architecture, so named because of its four principal registers, was invented by Landin [38] and described in detail by Henderson [28]. Each of the four registers is referred to as containing an S-expression, which in a real implementation will be a pointer to a data structure representing the expression. The term "stack" is used to refer to the data structure "within" a register. The four registers are:

S	*stack*	holds intermediate computation results
E	*environment*	holds values bound to variables during evaluation
C	*control list*	holds the machine-language program being executed
D	*dump*	saves values of other registers on calling a new function.

It is important to note that the entire state of the machine can be denoted by giving the content of its four registers. Instructions are defined by state transitions, enabling an interpreter to be developed by pattern matching, and the use of structural induction in proofs about the machine execution.

Each instruction is defined in terms of its effect on the machine state. For example, the ADD instruction definition is:

$$(a\ b.s)\ e\ (ADD.c)\ d\ \rightarrow\ (b+a.s)\ e\ c\ d$$

This instruction expects two arguments on top of the S stack. After execution, the two arguments are replaced by their sum, which is *cons*'ed onto whatever was below the arguments on the stack. The E and D registers are unaltered, but the C stack contains the rest of the control list that followed the ADD instruction.

Similar to the interpreter presented previously, a means of storing bindings is provided by a valuelist stored in the environment stack E, which is, as before, a list of lists. Instead of an associated namelist, variable names are replaced with a LD instruction and a pair of arguments telling it where to find the value in the environment. The first determines which list and the second which element from that list to retrieve. Two auxiliary functions, *index* and *locate*, are defined:

$$index(n,s) = if\ (n{=}0)\ then\ (car(s))\ else\ (index(n{-}1,\ cdr(s)))$$
$$locate(i,e) = index\ (cdr(i),\ index(car(i),e))$$

For example, the namelist *((x y)(w z))* would be translated as position indices *(((0.0)(0.1))((1.0)(1.1)))*. Thus the transition for LD in Table 2.3 shows that the value loaded on the stack is the value retrieved by applying the function *locate* to the arguments and the environment list.

Loading a constant is achieved by the LDC instruction, which takes an in-line constant as an argument, and places it on top of the stack.

All the arithmetic instructions work similarly to the ADD instruction described earlier: the stack is expected to have two arguments on top and they are replaced with the value obtained from applying the arithmetic operation to them. Notice the order of arguments for the noncommutative arithmetic operations SUB, DIV, REM and LEQ. The instructions effecting the structural operations CAR, CDR, CONS, and ATOM similarly expect suitable values on top of the stack, and replace them with the result of applying the primitive structural operation.

Two commands are used to implement the conditional branch. The SEL command expects a boolean value on top of the stack S, and two code sequences as arguments. If the value on the stack is T, then the first argument is installed in the control register C, otherwise the second argument is so installed. In either case, the remainder of the

	INITIAL STATE				TRANSFORMED STATE			
S	**E**	**C**	**D**		**S**	**E**	**C**	**D**
s	e	(LDC x.c)	d	→	(x.s)	e	c	d
s	e	(LD (m.n).c)	d	→	(x.s)	e	c	d
					where x = locate ((m.n),e)			
(a b.s)	e	(ADD.c)	d	→	(b+a.s)	e	c	d
(a b.s)	e	(SUB.c)	d	→	(b−a.s)	e	c	d
(a b.s)	e	(MUL.c)	d	→	(b×a.s)	e	c	d
(a b.s)	e	(DIV.c)	d	→	(b/a.s)	e	c	d
(a b.s)	e	(REM.c)	d	→	(b rem a.s)	e	c	d
(a b.s)	e	(EQ.c)	d	→	(b=a.s)	e	c	d
(a b.s)	e	(LEQ.c)	d	→	(b$\leq$a.s)	e	c	d
((a.b).s)	e	(CAR.c)	d	→	(a.s)	e	c	d
((a.b).s)	e	(CDR.c)	d	→	(b.s)	e	c	d
(a b.s)	e	(CONS.c)	d	→	((a.b).s)	e	c	d
(a.s)	e	(ATOM.c)	d	→	(t.s)	e	c	d
					where t = (a is an atom)			
(x.s)	e	(SEL c_t c_f,c)	d	→	s	e	c_x	(c.d)
					where c_x = if (x=T) then c_t else c_f			
s	e	(JOIN)	(c.d)	→	s	e	c	d
s	e	(LDF c'.c)	d	→	((c'.e).s)	e	c	d
((c' e')v.s)	e	(AP.c)	d	→	NIL	(v.e')	c'	(s e c.d)
((c' e')v.s)	(Ω.e)	(RAP.c)	d	→	NIL	rplaca(e',v)	c'	(s e c.d)
(x)	e'	(RTN)	(s e c.d)	→	(x.s)	e	c	d
s	e	(DUM.c)	d	→	s	(Ω.e)	c	d
s	e	(STOP)	d	→	s	e	(STOP)	d

Table 2.3: SECD Machine Instruction Definitions

control list after the two arguments is saved on the dump register D, to be restored once the selected branch control sequence is finished executing. That is the purpose of the JOIN command, which will be the last item in each of the two control list arguments to the SEL command. Thus, the execution of the code in either control sequence must leave the value stored in the D register untouched at the end of its execution.

The final set of instructions is used in implementing functions and function applications. In the interpreter for LispKit, functions are represented by a closure, containing the bound variables, the expression part of the function, and the context in which free variables are to be evaluated. In the SECD machine, the expression consists of a control list and the context is represented by an environment list. A function object is created by the LDF instruction. It takes a code sequence as an argument, and *cons*'es it onto the current environment, leaving this on top of the stack S. The function object is usually installed in an environment and recalled with an LD instruction to be applied to different arguments.

The AP instruction is used to apply a function to a list of parameters found immediately below the function object on top of the stack S. AP installs the parameters in the environment using the context part of the function object, loads the control part in the C register, empties the stack S, and saves the existing register contents on the D stack for later restoration by the RTN instruction. From inside the function code, parameters are accessed by selecting items in the first list in the environment, using position indices (0.0), (0.1), etc., while variables free to the body are accessed using position indices (1.0), (1.0), (2.0), etc. The RTN instruction expects the dump D to be unaltered by the execution of the control list, so that it can restore the state of the registers upon completion of the function call, with the value returned by the function call installed on top of the stack S.

Recursive functions are more complicated, since the environment must contain a copy of the function object itself, for recursive calls to the function. Thus, a circular data structure is necessary. As in the interpreter, a *rplaca* operation is used to replace the value representing the pending value of the environment (represented here by the symbol Ω) with the intended environment.

DUM creates the environment with the installed Ω placeholder. The RAP instruction is similar to the AP instruction, except that in creating the environment in which the function is executed, the parameters are installed by using the *rplaca* operation on an environment whose *car* is the pending value Ω. RAP will always be executed in a state where the environment part of the function value, e', is the same object as the contents of the environment register (Ω.e). The list of parameters must consist of only function objects, and they will all contain the same environment component, so the single destructive operation creates a circular context for all the mutually recursive definitions.

The final instruction, STOP, terminates computation.

2.4 LispKit to SECD Machine Code

The execution of LispKit programs on the SECD architecture requires their compilation into machine code. The compiled program, a function-valued object, loaded into the C register of the SECD machine, with an argument list loaded into the S register, should, upon completion of execution, leave a single result on the stack S, and that value should match the result of executing the LispKit program in the interpreter. The compiler will be defined by describing the translation of each well-formed LispKit expression.

The machine code is generated with respect to a namelist, which is built up as the expression is compiled. This namelist is used to keep track of where values associated with variables will be found in the environment at execution time. A function is defined to extract a position for any variable in the namelist.

> *location (x,n) =*
> *if (member(x, car(n))) then (cons(0, position(x,car(n))))*
> *else (cons(car(z)+1, cdr(z)))*
> *where z = location(x, cdr(n))*

Following the usage of Henderson, the definition in Table 3.2 makes use of the infix operator "|" to represent the *append* function.

The compilation of an expression with respect to a namelist n is represented by "$*n$".

LispKit expression	Compiled code
$x*n$	(LD i) where i = location(x,n)
$(QUOTE\ s)*n$	(LDC s)
$(ADD\ e_1\ e_2)*n$	e_1*n | e_2*n | (ADD)
$(SUB\ e_1\ e_2)*n$	e_1*n | e_2*n | (SUB)
$(MUL\ e_1\ e_2)*n$	e_1*n | e_2*n | (MUL)
$(DIV\ e_1\ e_2)*n$	e_1*n | e_2*n | (DIV)
$(REM\ e_1\ e_2)*n$	e_1*n | e_2*n | (REM)
$(EQ\ e_1\ e_2)*n$	e_1*n | e_2*n | (EQ)
$(LEQ\ e_1\ e_2)*n$	e_1*n | e_2*n | (LEQ)
$(CAR\ e)*n$	$e*n$ | (CAR)
$(CDR\ e)*n$	$e*n$ | (CDR)
$(CONS\ e_1\ e_2)*n$	e_2*n | e_1*n | (CONS)
$(ATOM\ e)*n$	$e*n$ | (ATOM)
$(IF\ e\ e_1\ e_2)*n$	$e*n$ | (SEL e_1*n | (JOIN) e_2*n | (JOIN))
$(LAMBDA\ (x_1\ ...x_k)\ e)*n$	(LDF $e*((x_1\ ...x_k).n)$ | (RTN))
$(e\ e_1\ ...e_k)*n$	(LDC NIL) | e_k*n | (CONS) | ... | e_1*n | (CONS) | $e*n$ | (AP)
$(LET\ e\ (x_1.e_1)\ ...(x_k.e_k))*n$	(LDC NIL) | e_k*n | (CONS) | ... | e_1*n | (CONS) | (LDF $e*m$ | (RTN) AP) where $m = ((x_1\ ...x_k).n)$
$(LETREC\ e\ (x_1.e_1)\ ...(x_k.e_k))*n$	(DUM LDC NIL) | e_k*m | (CONS) | ... | e_1*m | (CONS) | (LDF $e*m$ | (RTN) RAP) where $m = ((x_1\ ...x_k).n)$

Table 2.4: SECD Code Generated for Well-Formed Expressions

A simple **variable** is compiled into code that loads the value from the environment list that is located in the corresponding location to the variable in the namelist. A **constant** value becomes a LDC instruction with the constant as argument.

An **arithmetic expression** compiles to a code sequence that first loads the values of the arguments, followed by the appropriate machine code instruction to execute the operation. The order of arguments is easily understood if we expect the LispKit expression $(SUB\ e_1\ e_2)$ to evaluate to $e_1 - e_2$. The code generated will cause the value of e_1 to be loaded on the top of the stack, and then the value of e_2 will be placed on top of that. Thus the stack looks like $(val(e_2)\ val(e_1).s)$, which clarifies why the machine transition definitions consider the top of stack as the second argument to the arithmetic operation. Notice that the *CONS* instruction is just the opposite, and the code sequence for the arguments is reversed in the resulting code sequence.

The **conditional expression** is compiled to a code sequence that first loads the value of the conditional part of the expression, followed by a SEL instruction, and this in turn is followed by the code sequences for each of the two branches of the conditional, both of which end with a JOIN instruction.

A **function** defined by the use of *LAMBDA* is compiled into a control sequence that consists of a LDF instruction followed by the code for the function body compiled with respect to the namelist augmented with the list of locally bound variables, and with a RTN instruction appended to the end. The LDF instruction takes the control list for the function body as its argument, and creates a closure at run time by *cons*'ing this onto the current environment.

The code generated for a **function application** will first build a list of values for each of the arguments to the function on top of the stack S. This is followed by the code for the function object being applied. This could consist of a nameless *LAMBDA* expression generating the code just seen, or the name of the function, which would also cause the function closure to be loaded on top of the stack with a LD instruction. The last instruction is AP, which will effect the application of the function to the arguments placed on the stack. It should be noted that a *LET* expression compiles into precisely the same code sequence as an application of a *LAMBDA* expression.

The final well-formed LispKit expression is the **recursive block**. An initial DUM instruction modifies the current environment by installing a placeholding Ω value. This is followed by a code sequence that builds a list of the values to be bound to each of the local variables. Each of these is expected to be a function-valued object, and will thus contain a copy of the environment with the placeholding Ω value in its closure. The expression is treated as a function, with a LDF instruction and the code sequence for the expression followed by a RTN instruction. The RAP instruction is used instead of the AP instruction to create the required circular environment structure.

Henderson [28] gives a complete LispKit to SECD code compiler written in LispKit.

Example 2.3 Compiling *(twice double)*

We use the same nonrecursive example function used earlier. The compilation steps are presented by showing the code skeletons generated for each recursive call to the compiler.

(LET (twice double)
  (double LAMBDA (x) (ADD x x))
  (twice LAMBDA (f) (LAMBDA (x) (f (f x)))))*()
$\Longrightarrow$ (LDC NIL) |
  (LAMBDA (f) (LAMBDA (x) (f (f x))))*() |
  (CONS) |
  (LAMBDA (x) (ADD x x))*() |
  (CONS) |
  (LDF *(twice double)*((double twice))* | (RTN)
  AP)

(LAMBDA (f) (LAMBDA (x) (f (f x))))*()
$\Longrightarrow$ (LDF *(LAMBDA (x)(f (f x))*((f))* | (RTN))

(LAMBDA (x)(f (f x)))*((f))
$\Longrightarrow$ (LDF *(f (f x))*((x)(f))* | (RTN))

(f (f x))*((x)(f))
$\Longrightarrow$ (LDC NIL) | *(f x)*((x)(f))* | (CONS) | *f*((x)(f))* | (AP)

(f x)*((x)(f))
$\Longrightarrow$ (LDC NIL) | *x*((x)(f))* | (CONS) | *f*((x)(f))* | (AP)

x*((x)(f))
$\Longrightarrow$ (LD (0 . 0))

f*((x)(f))
$\Longrightarrow$ (LD (1 . 0))

(LAMBDA (x) (ADD x x))*()
$\Longrightarrow$ (LDF (LD (0 . 0) LD (0 . 0) ADD RTN))

(twice double)*((double twice))
$\Longrightarrow$ (LDC NIL LD (0 . 0) CONS LD (0 . 1) AP)

The fully compiled function follows. Notice the appending of **(AP STOP)** to the end of the code. This is added to the end of the code list to be installed in the C register, to cause the function to be applied to the arguments installed in the E register.

```
(LDC NIL
 LDF
  (LDF (LDC NIL LDC NIL LD (0 . 0) CONS LD (1 . 0) AP CONS
  LD (1 . 0) AP RTN)
  RTN)
 CONS LDF (LD (0 . 0) LD (0 . 0) ADD RTN)
 CONS LDF (LDC NIL LD (0 . 0) CONS LD (0 . 1) AP RTN)
 AP AP STOP)
```

Loading this code into the C register of an SECD machine, with the S register containing the argument list (7), will eventually cause the machine to reach a state (28) () (STOP) (), after executing a total of 34 primitive instructions. The interested reader may care to follow the execution by writing a simple machine simulator which prints out the internal state after the execution of each machine instruction. Be warned that for recursive definitions, the function printing the internal state should not try to print all of the infinite (circular) data structures!

2.5   Summary

This chapter has described the LispKit language, and used it to illustrate the operation of the SECD architecture transitions. The LispKit syntax was given, and an informal semantics provided in the form of an interpreter function. The abstract SECD architecture was defined by giving the set of SECD language instructions, and defining state transitions for each instruction. A translation schema for LispKit into SECD code was described, and its implementation as a compiler written in LispKit supplied. This translation along with the SECD definition provides an operational semantics for LispKit. A proof of correctness of the translation and a full description of the semantics of LispKit may be found in [51]. While not central to this work, the introduction of a higher level language demonstrates the operation and potential of the SECD architecture. More extensive programming examples in LispKit are available, for example [29, 30]. Furthermore, the simple translation from the higher level language to SECD code demonstrates the suitability of the architecture for executing functional programs.

The following chapter proceeds to develop the architecture in greater detail, coming eventually to the design of a working hardware system.

Chapter 3

SECD Architecture: Silicon Synthesis

With the informal definition of the SECD machine as a starting point, this chapter follows the design of the hardware system by progressive transformation of the original specification. We focus on two major themes: development of the external architecture (the machine as seen by its users), largely influenced by external constraints and decisions, and development of the internal architecture (how the machine is physically organized), through a progressive elaboration of the system model. Both are developed within a framework of increasingly detailed levels of description of the system.

The implementation does not aspire to state of the art for hardware design, but is interesting as a study of transforming a specification into a working system. The detailed description of the system will be important in appreciating some of the subtleties of the formal specification and verification to follow.

3.1 Project Context

The SECD chip arose within a larger ongoing research effort by the VLSI group at the University of Calgary. The chip was used as a vehicle to explore the use of specification to drive design synthesis. The methodology entails elaborating a design hierarchically as a tree of nodes and formally specifying the behaviour at each node. Verifying that the composition of behaviours of a node's children agrees with the node's specification assures a correct design. By deductive argument, the correctness or otherwise of a complete design can be shown. While the chip provided the team with hands-on experience with a nontrivial design, the focus of study was the design process, and this strongly influenced the dominant design criteria.

- The most important criterion was that a *correct, working device* be produced. Correctness is the primary objective of the specification-driven process.

- The next criterion was *simplicity.* Simplicity was necessary on two counts: to ensure that verification could cope with what promised to be the most complex microprocessor proof attempted to date, and secondly, to improve the likelihood of meeting the first criterion.

- *Testability* of the design was considered essential. In the event of malfunction, determination of the source of the error requires examining the state of the machine extensively. Furthermore, correct output from test problems does little to assure total design correctness. Rather, each step of the computation should be accessible for examination.

- Lastly, *utility* should be considered. It was preferable that the design could be given tractable problems, rather than be considered a toy device, incapable of all but the most trivial tasks. A particularly relevant problem would be compiling LispKit programs to SECD code.

Equal in importance to the selection of criteria is the explicit statement of items that are not given priority. *Speed* was specifically eliminated as a determining criterion, both in terms of clocking rate, and optimality of the operation sequences, insofar as they could conflict with the simplicity criterion.

3.2   Levels of the Design

The wide gulfs between the external architecture view, captured abstractly in the top level specification, the internal architecture, expressed as an assembly of logical devices, and the layout, expressed as a set of masks, are most easily bridged by a succession of levels of description, with detail increasing at each lower level. Associated with each level is an interpreter expressed as a simulation model.

The simulation model has both a control structure, and a set of operations. The operations may be further expanded at the next lower level. There were seven major major levels of description used in the development of the SECD chip.

Abstract machine is the high level definition of the machine characterised by the contents of 4 stacks, defined by the state transitions of Table 2.3.

Abstract System level views the SECD machine as a batch mode co-processor, with external 'read' and 'print' routines to download problems and return results. The decomposition of transitions into operation sequences begins here.

Top level FSM (finite state machine) describes the control of the system in terms of major states and transitions, introducing control inputs for initialisation and state transition.

Abstract RTL (register transfer level) view begins elaborating the internal architecture, adding registers, combinational logic devices, the bus, and memory, and determining data representation and word configuration.

Concrete RTL develops the control part of the internal architecture as a microcode program[1].

Mossim defines the design down to the transistor level, and determines the design of memory elements and the clocking scheme.

Layout level generates a set of masks, after resolving floor planning, electrical and other low level concerns.

The first 3 levels of description mainly develop the external architecture, while the remaining levels are concerned with developing the internal architecture. We shall give an overview of major aspects of the design, without dwelling at length on minor detail, or describing particular simulation models in full.

---

[1]The microcode for the SECD microprocessor was written by J. Joyce [33].

3.3   The Chip Interface

The transition relation shown in Table 2.3 defined over the machine instructions provides a semantics for the machine language, but as a machine specification is somewhat inadequate. Realising the SECD machine as a working system required that it be able to accept a task and compute and return a result. The interface to permit a user to pose a problem and the machine to return a result was the first major design decision[2].

Using the SECD as a co-processor to another system would permit i/o to be handled by the other system rather than the SECD chip. For instance, using SECD as a co-processor for a SUN workstation would see the SUN able to read in an S-expression, set up a memory image for the problem, signal the SECD to begin computation, receive a signal back on completion of the calculation, and print out the S-expression solution. This has the advantage of simplifying the tasks the SECD must perform.

On the other hand, designing a stand-alone system would require incorporating primitive read and write operations into the definition of the machine. Ideally, an operating system for such a system would be written in the higher level LispKit language, but its pure functional nature provides recursion as the sole means to achieve the infinite 'while' loop required, but since every recursion uses up system resources, this would exhaust memory. A possible solution would be to hand modify the machine code to create the desired loop.

An extension of the stand-alone system concept is the multi-processor SECD system. Such a system was envisioned as having multiple (perhaps 100) SECD chips operating concurrently on shared memory. An operating system kernel would assign S-expressions to processors for evaluation. Each processor would be assigned exclusive use of memory blocks in a sort of virtual memory system, with blocks being assigned and garbage collected by the kernel.

It is not surprising, in light of the focus on formal verification and the scope of the overall project at the time, that the co-processor option was chosen.

---

[2]This discussion summarises work by Jeff Joyce on system configuration, reported originally in [33].

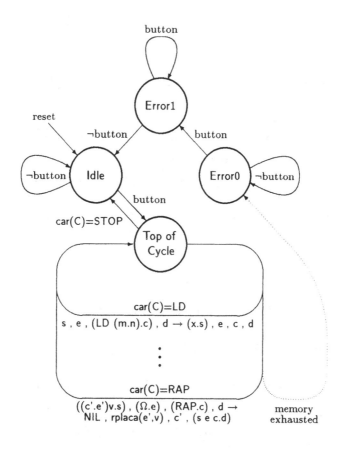

button

Error1

reset

¬button   button

¬button Idle   Error0 ¬button

car(C)=STOP   button

Top of
Cycle

car(C)=LD

s , e , (LD (m.n).c) , d → (x.s) , e , c , d

.
.
.

car(C)=RAP

((c'.e')v.s) , (Ω.e) , (RAP.c) , d →   memory
NIL , rplaca(e',v) , c' , (s e c.d)   exhausted

Figure 3.1: Top Level Finite State Machine View of SECD

A four state finite state machine model expresses the desired operating interface. Several ideas are incorporated including the addition
of a *reset* input to permit a deterministic startup of the machine, an
idle state for waiting until a problem is loaded, a *top_of_cycle* state
as the top of a fetch/execute cycle, two *error* states, and a *button*
input to control transitions from *error* and *idle* states. This view
is summarised by Figure 3.1. The inclusion of error states recog-
nizes that computations may require in excess of the available finite
memory space, and thus failure can occur. The sharing of the *button*
input for both startup and error recovery transitions motivates the

use of two distinct error states which makes the input signal timing less critical.

3.4   Internal Architecture and Microcode

Henderson [28] pointed out that it is possible to derive an interpreter for the SECD machine by noting how individual instructions update the (S, E, C, D) state 4-tuple as they are executed. When designing a control sequence for the interpreter, a precedence (partial ordering) was established on registers for each machine instruction to prevent overwriting. As an example, consider generating a control sequence for the AP instruction transition. The abstract machine transition is given by:

$$((c'.e')v.s) \quad e \quad (AP.c) \quad d \quad \rightarrow \quad NIL \quad (v.e') \quad c' \quad (s\ e\ c.d)$$

from which the following precedence on registers is observed:

$$D \prec \frac{E}{C} \prec S.$$

The control sequence is then derived by pattern matching.

$$D := (cons\ (cdr\ (cdr\ S))\ (cons\ E\ (cons\ (cdr\ C)\ D)));$$
$$E := (cons\ (car\ (cdr\ S))\ (cdr\ (car\ S)));$$
$$C := (car\ (car\ S));$$
$$S := NIL;$$

This control sequence will form the basis for a microcode program, once fundamental decisions about the internal architecture are settled.

3.4.1   Data representation

S-expression data structures must be represented within a memory. Three types of objects need to be represented: numbers, symbols and cons (dotted pair) records. Two bits of each record will be assigned to identify its type. A simple mark and sweep garbage collector requires the use of two additional bits in each record. The final word configuration follows.

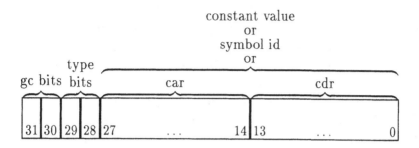

Two's complement representation is used for numbers, which can range over the representable integers. Symbols represent atomic values which can only be tested for equality with each other. Thus, a distinct symbol identification number is a suitable representation. The "meaning" of the symbol (or its written form) is of concern only on input and output operations, and hence assignment and interpretation can be handled entirely outside the SECD chip (by the compiler, since new symbols cannot be created in the course of executing programs). Three symbolic constants (NIL, TRUE, and FALSE) are required by the SECD chip, and will be "built-in" as constants.

Cons records represent a pairing operation of S-expressions. In typical Lisp fashion, these are implemented by pairs of pointers to other cells. The size of pointer determines the memory address space, and hence the maximum problem size that the machine can compute. As a minimum, the SECD machine should be able to run the LispKit compiler on the compiler program itself, and this required approximately 2^{12} words. This sets a lower bound of a 28 bit word $((2 \times 12) + 4)$. The availability of memories in multiples of 8 bits makes 32 bit words an appealing choice.

3.4.2 The register level view

We now refine the top level view into a concrete architecture. For ease of implementation and verification, we use a single bus architecture and an external RAM memory. The S, E, C, and D stacks are implemented as 14 bit registers that contain pointers to S-expressions in memory. The free list, used to allocate unused cells as required by the computation process, is similarly implemented by a register labelled *free*, holding a pointer to the free list in memory. Working registers *x1* and *x2* are added to permit computation of intermediate results as arguments to a *cons* operation. A 32 bit *arg* register is added to hold integer or symbol arguments for alu operations, and generally for holding 32 bit records read from memory, including the machine instruction codes. The output of the *alu* is connected to two 32 bit buffer registers. The 32 bit *buf1* register is used to store the alu output during normal operations, while the *buf2* register is used similarly during garbage collection.

With each addition to the hardware, a functionality or role is determined, and thereafter this functionality is respected. Nontransparent uses of components is avoided, with the expectation that this will make the verification task more manageable. The clearest indication of this approach is the provision of separate registers: *root, parent, y1,* and *y2* (in addition to *buf2* mentioned above) for use by the garbage collector.

We can now systematically generate an register-register level microcode program from the control sequences. Since *car*, *cdr*, and *cons* operations are very common we anticipate special hardware for them. At this stage we mechanically rewrite the operations for AP into smaller steps, each step being of the form $R := rhs;$ where R is a register and *rhs* is either a register, or a single *car*, *cdr* or *cons* operation on a register, or a primitive alu operation on the *arg* plus one other register for binary operations.

Much of the time we are assembling sub-expressions in order to cons them together. We simplify matters by using the two general purpose temporary registers *x1* and *x2*. When we are carrying out a *cons* operation, we put the second argument in *x2* and the first argument in *x1*.

Two combinational logic elements are assumed at this stage. The alu primary function is the computation of values for the arithmetic SECD machine operations: **ADD, SUB, MUL, DIV,** and **REM.** Additional operations are required for the garbage collector, including setting and clearing of the mark and field bits, and the destructive *replcar* and *replcdr* operations used for the in-place traversal of the data structures in memory. Lastly, there is a *decrement* operation, used in looking up values in the environment. It is also used in the "sweep" phase of garbage collecting to step through the mem-

```
! D := (cons (cdr (cdr S)) (cons E (cons (cdr C) D)));
x2   :=   D;
x1   :=   cdr C;
x2   :=   cons x1 x2;
x1   :=   E;
x2   :=   cons x1 x2;
x1   :=   cdr S;
x1   :=   cdr x1;
D    :=   cons x1 x2;
! E := (cons (car (cdr S)) (cdr (car S)));
x2   :=   car S;
x2   :=   cdr x2;
x1   :=   cdr S;
x1   :=   car x1;
E    :=   cons x1 x2;
! C := (car (car S));
C    :=   car S;
C    :=   car C;
! S := NIL;
S    :=   NIL;
```

Table 3.1: Initial microcode sequence for the **AP** instruction

ory address space. The high address value is built-in as a constant to provide a starting point for the sweep. These two uses have distinct data type arguments: the first uses integers, while the second is applied to addresses. Thus, the 14 bit addresses had to be padded out with zeros to make 28 bit integers. For this purpose, a constant register *(the clearunit)* loads zeros onto the upper 14 bits of the bus when required. This is the only place in the design that blatantly violates the principle of distinct components for distinct functions.

The flagsunit returns the boolean result of predicates used both for the control of the *if ... then ... else* and the *while* structures, as well as computing the SECD machine operations **EQ** and **LEQ**. With the addition of an *mar* register to select locations for memory access, the architecture is now complete. Figure 3.2 summarises the operational part of the architecture at this level, and pictures the control part as a classic finite state machine.

3.4.3   The microcode program

The next level of refinement concentrates on transforming the control sequence into the final series of microcode instructions. The transformation is accomplished in several steps, and the final resulting sequence can then be compiled into a binary image to automatically generate a microcode ROM.

A microcode sequence is generated by mechanically translating each of the four higher level functions into an instruction sequence. In the main, each register-register operation is split into a read onto the bus and a write from the bus. Each register X has two control signals associated with it, systematically called rX and wX. When rX goes high, the contents of X are read onto the bus. The controller sees to it that only one register may be read onto the bus at a time and for simplicity, only one register may be written from the bus at a time. When wX goes high, the contents of the bus are written into X: all the bits of the bus if X is a 32-bit register; if X is a 14-bit register, the top 18-bits of the bus are ignored except for the special CAR register which is wired to accept bits 14-27 of the bus. Bread and butter microcode can be generated directly from the register-register level description as soon as we supply a template for each class of assignment in the register-register level description.

- Copying the contents of register X to register Y is quite simply

$$rX \qquad wY$$

Movement involving *cdr*, *car* or *cons* require access to the external memory, and a memory address register *(mar)* is introduced to select memory locations.

- To carry out the request $Y := cdr\ X$ we place the contents of X in *mar*, read *memory[mar]* onto the bus (all 32 bits), and then store its *cdr* field in Y.

$$rX \qquad wmar$$
$$rmem \qquad wY$$

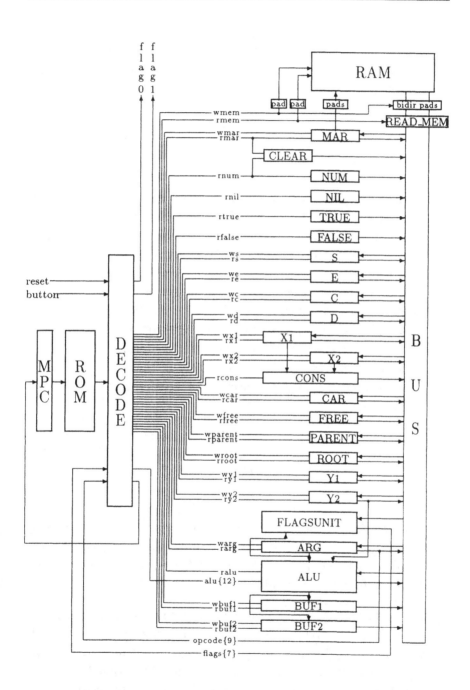

Figure 3.2: Register Transfer Level View of SECD Machine

- To carry out the request $Y := car\ X$ we place the contents of X in *mar*, read *memory[mar]* onto the bus (all 32 bits), then filter out the *car* bits via the *car* register before passing the result to Y.

rX	wmar
rmem	wcar
rcar	wY

- The much used *cons* operation requires more thought. A call on *cons* requires the freeing of a fresh cell in memory (perhaps initiating a garbage collection) and the passing of two addresses to memory to fill out its fields. We restrict *cons* to operating upon two specific registers *x1* and *x2*. Our *consx1x2* operation locates a free cell in memory and updates the *free* register, fills out the two address fields of the cell from *x1* and *x2*, sets its type bits to indicate a cons'ed word, zeroes its garbage collection bits, writes it to the memory, and returns its address. Calls to *consx1x2* occur so frequently in the code and are of such complexity that we insert a subroutine facility into the microcode.

if (null rfree) then GCBegin()	
rfree	wmar
rmem	wfree
rcons	wmem

Notice that the subroutine returns the address of the cons cell in the *mar* register. The return address from the subroutine will be included as an argument to the call, and a stack will be needed to store these addresses.

The last aspect of the microcode design concerns flow of control during execution. Five mechanisms are recognized: unconditional jumps, conditional jumps, subroutine calls, subroutine returns, and a jump table that uses the current machine instruction value. Of these, the conditional jumps and subroutine calls both required two addresses, one of which will always be the following instruction's address, while the others required a single address argument. The simple sequencing forward to the next instruction is interpreted as an unconditional jump as well. The subroutine mechanism requires a microaddress stack. Eight conditional jump instructions have conditions:

- the value of the *button* input

- is the argument an *atom* record?

- the EQ operation applied to 2 arguments

- the LEQ operation applied to 2 arguments

- is the record equal to the symbolic constant NIL?

- is the record equal to the symbolic constant TRUE?

- is the *mark* bit set?

- is the *field* bit set?

These values are the flagsunit outputs. Additionally, the current machine instruction value is needed by the controller. The controller thus has 12 distinct ways of selecting the next microinstruction address.

Relying on mechanical rewriting to complete the transformation to microcode misses obvious optimizations, such as not reloading the same value into a register if it can be established that it is already present. A peephole optimizer can be effective at eliminating many of these redundant command sequences. Recognizing the similarity of most of the steps of the arithmetic instruction sequences invites the addition of another level of subroutine calls to share code. The unconditional jump instruction was divided into remote jumps and jumps to the following microinstruction, with the latter not needing a specified argument. This increased the number of ways of controlling flow of command in the control unit to 13. The final version of the microcode for the AP instructions is shown in Table 3.2.

This microcode program is compiled to a binary image and then fed into a ROM layout generator.

Our modelling of the chip implicitly assumes an external RAM, since an outside agency is expected to download problems and upload

```
L("AP");
/* D */
          rd      ; wx2             ; (inc ()) ;
          rc      ; wmar            ; (inc ()) ;
          rmem    ; wx1             ; (call ("Consx1x2")) ;
          rmar    ; wx2             ; (inc ()) ;
          re      ; wx1             ; (call ("Consx1x2")) ;
          rmar    ; wx2             ; (inc ()) ;
          rs      ; wmar            ; (inc ()) ;
          rmem    ; wx1             ; (inc ()) ;
          rx1     ; wmar            ; (inc ()) ;
          rmem    ; wx1             ; (call ("Consx1x2")) ;
          rmar    ; wd              ; (inc ()) ;
/* E */
          rs      ; wmar            ; (inc ()) ;
          rmem    ; wcar            ; (inc ()) ;
          rcar    ; wmar            ; (inc ()) ;
          rmem    ; wx2             ; (inc ()) ;
          rs      ; wmar            ; (inc ()) ;
          rmem    ; wx1             ; (inc ()) ;
          rx1     ; wmar            ; (inc ()) ;
          rmem    ; wcar            ; (inc ()) ;
          rcar    ; wx1             ; (call ("Consx1x2")) ;
          rmar    ; we              ; (inc ()) ;
/* C */
          rs      ; wmar            ; (inc ()) ;
          rmem    ; wcar            ; (inc ()) ;
          rcar    ; wmar            ; (inc ()) ;
          rmem    ; wcar            ; (inc ()) ;
          rcar    ; wc              ; (inc ()) ;
/* S */
          rnil    ; ws              ; (jump ("top_of_cycle")) ;
```

Table 3.2: Final Microcode Sequence for **AP** Instruction

results, and there is no provision in the model for handing control of the memory to the external agency. External RAM is consistent with the simplicity criterion, and focussed our effort on the microprocessor design, rather than the distinct concerns of RAM design. The RAM is treated as just another, though addressable, register, with read and write signals controlling it. It is expected that the RAM will default to a read operation, and the rmem control line will control its gating onto the bus. The high memory address constant used by the garbage collector will also be used to reference a reserved memory location used to hold a pointer to a downloaded problem in memory, and to hold a pointer to the result on completion of computation.

This classical finite state machine implementation of the controller, with the state held in the **mpc** register and extended to include the microinstruction stack registers as well, changes with each microcode instruction executed.

3.4.4 SECD layout

The register transfer level defines both a control part (the microcode) and an operative part (the registers, logical units, bus, and memory). The given operations can be completed within a clock cycle, and indeed, the time quantum of the model matches the clock frequency. Next we model the SECD design down to the transistor level, where the design of the memory devices and the clocking scheme are the major issues. The implementation will be in a CMOS technology.

Previous views of the SECD divided it into two major functional components: the controller and the datapath. They have been developed somewhat independently, and making them independently testable seems appropriate. If a flaw were to occur in one component, testing of the other component would still be possible. To meet this objective, and previously stated testability concerns, we shall add a bank of shift registers between the two components, which can be used to trap all, or most, signals passing between them. The shift registers should be transparent during normal operation, and only during test mode will they be used to shift out an internal state vector.

Memory Elements and Clocking

Level-triggered latches are selected largely for reasons of space and transistor count efficiency. Level triggered latches are also in keeping with the view of circuits presented in by Mead and Conway [42] as a system of opening and closing valves. In some sense, level triggered latches can be viewed as falling edge triggered, although the previous state is lost at the start of the clock pulse.

Implementing the controller as a finite state machine requires buffering between current and next states. This is achieved by the use of a two-phase non-overlapping clocking scheme and paired master/slave registers, along the design style described in [42]. The state register in the control unit is the **mpc** register, but in a more general sense, the values on the 4-deep microcode subroutine stack are also part of the state. In the following discussion, references to the **mpc** register can be applied similarly to the stack registers. The **mpc** register is actually the slave register, the master is labelled **nextmpc**. **Nextmpc** is clocked on the ϕ_A phase, and **mpc** on the ϕ_B phase. The control unit state is considered to change on ϕ_B. Using the terminology of Anceau [1], this arrangement for synchronization of the circuit uses "mixed polyphase mode", and will be stable if a wait state is present in the clock cycle. The short time required to pass values from *nextmpc* to *mpc* along with a sufficiently low frequency ensures a wait before ϕ_B rises.

Particular attention must be paid to possible *race* conditions. One example is the possibility of generating transient "write" signals from the ROM when the value **mpc** is changing. One solution is to delay latching of the datapath registers until after the **mpc** is latched (and the value propagated). Use of the inverse clock signal $(\overline{\phi_B})$ is still subject to race conditions[3], so the ϕ_A phase is used instead. The clocking scheme requires that inputs to registers latching on ϕ_A be stable prior to the end of the ϕ_A pulse. The overall view of the chip now sees the control unit as changing state on ϕ_B and the datapath changing on ϕ_A.

The interface to the external memory gives rise to the possibility of undesirable power dissipation. This can occur if both memory and SECD devices drive external lines simultaneously for some overlapping interval, which is possible given capacitance induced delays of signal switching. A more considered interface than the "memory as a register" view used until now is needed.

---

[3]One could devise a scenario where the $\overline{\phi_B}$ signal overlaps the start of the ϕ_B pulse, and thus random write signals may be generated.

Bidirectional i/o pads are used to connect the data bus to the external RAM. The default mode for the pads is input, switching to output mode only when writing to memory. A set of busgates isolates the input from the datapath bus unless a read memory operation is signalled. A critical race occurs when a *write* immediately follows a *read* from memory (this occurs 7 times in the final microcode). The problem is that a typical static RAM memory chip may still drive its output after the external *rmem* signal is not asserted[4]. It is possible that the *wmem* signal may be asserted, thus causing the bus to drive off-chip, before the memory chip has stopped driving its outputs.

The solution involves both delaying the enabling of the bidirectional pads during a write operation, and shortening the off-chip *wmem* pulse. The first is achieved by NAND'ing the *wmem* signal with $\overline{\phi_B}$ (the active low generated signal is labelled *wrt-en*). The second is achieved by NAND'ing the same *wmem* signal with ϕ_A. By ensuring the interval between the clock pulses is greater than the chosen RAM chip's driven delay, we eliminate the possibility of harmful power dissipation.

The shift registers which separate the controller from the datapath will use the same level-triggered, paired latch design, but will be supplied with a separate pair of clock signals, labelled sr-ϕ_A and sr-ϕ_B. Aside from a input *shin* and output *shout*, two control signals are needed. *Shift* controls the input selection for each shift register cell, selecting either the internal line input or the previous cell output. *Test* selects the signal passed back to the chip; either the value held in the shift register cell, or the internal line of the chip, bypassing the shift register entirely (this is the normal mode of operation for the chip).

Control Unit

The final microcode has 399 instructions, and if a fully horizontal encoding is used, will require 74 bits. In order to reduce the size of the ROM, and encoding scheme is adopted. The individual sets of

---

[4]Typical delay times of up to 50 nanoseconds for outputs still driven after read signal is deasserted are given for the AM99C88/AM99CL88 CMOS Static RAM memory chips.

read, write, alu control, and *test*[5] signals (23, 17, 12, and 13 signals respectively) are mutually exclusive, and thus can be readily encoded. Encoding the signal sets separately permits easier examination for error detection, and will also provide a simple encoding for later verification. This reduces the microinstruction word length to 27. The 9 bit address field, though sparsely used, is retained in every instruction.

The layout of the ROM itself uses a 7 bit row decode and 2 bit column decode, providing a nearly square device, allowing considerable flexibility in the control unit layout. The ROM layout is generated automatically from a bit pattern produced from the microcode program. The decode component is a fully complementary CMOS device, while the 'OR' is implemented in a pseudo NMOS design, using pull-up transistors in each column, and n-type devices exclusively in the plane.

Having the ROM output encoded signals necessitates decoders and the same automated ROM/PLA generator is used to produce the layout of all 3 decoders. The 13 alternative methods of selecting the next microcode instruction select from only 4 possible values:

- the address following the current one in the microcode,

- the address supplied as the A field in the microcode,

- the SECD machine instruction code (opcode), and

- the value on top of the microcode subroutine stack.

A 4 × 1 MUX gates these values to nextmpc register. The mux control signals are generated by a *test* field of the microcode, in combination with the value of the *flag* and *button* inputs. The logic is implemented in a PLA, again generated automatically.

A bit-slice approach is used for the **mpc** and **nextmpc** registers, the microcode stack, and the logic to implement the selection of the next microinstruction. The required control signals for this are generated from a single row of random logic. A bit of random logic is required for other signals, such as the control for the bidirectional

---

[5] *Test* signals determine flow of control in the microcode.

i/o pads. Two output signals, *flag0* and *flag1*, are added to identify three of the major states (the *error1* state is not distinguished).

Datapath

The datapath is designed around a 32 bit bus, connecting all the registers and combinational logic devices. Most registers are simply 14 bits, connected to the *cdr* field of the bus, aside from the alu output **buffer** registers and the **arg** register which are all 32 bits. The **car** register inputs are connected to the upper 14 bit address field of the bus, while its outputs are connected to the lower (*cdr*) field. The **x1** register inputs and outputs both connect to the *cdr* field of the bus, but the output additionally connects to the *car* field inputs of the **consunit**. The **clearunit** sets the upper (*car*) field of the bus to zeros when the **mar** or **num** registers are read, because these are the sources of addresses that are decremented by the **alu**. This operation effectively maps a 14 bit address to a 28 bit integer. The **alu** is simplified by omitting the three most complex (in terms of area) operations: mul, div, and rem. Their opcodes were not eliminated however, and the implementation defaults to the dec operation.

Registers are grouped into subcomponents: **regs-14-misc, regs-14-car, regs-14-y2, regs-32-arg**, and **regs-32-bufs**. No error checking of type bits is implemented for any operations in the datapath. Logical **alu** operations maintain the (unaffected) bits, while the arithmetic operations produce 32 bit output with the type bits set to integer, and both mark and field bits cleared. Similarly, the **consunit** outputs a 32 bit value with record bits set to *cons* and mark and field bits cleared.

3.5 The Final Layout

The team working on the project included individuals with the expertise required for full custom design, and full custom CMOS fabrication was available through MOSIS, with a clear and concise set of scalable design rules. The layout design was completed using the 'Electric' system ([48]). A lengthy delay between the completion of the layout (late 1986) and the chip fabrication (autumn 1988) permitted a switch from the intended 3μ process to a 2μ process introduced by MOSIS in the interim. The use of scalable design rules was vital in taking advantage of the new technology *without any redesign.*

The major functional components, namely the datapath and controller, from the register transfer level were maintained as major floor plan elements. The shift register block was located between these two, and all were surrounded by a padframe.

A cell library was built, with guidelines rigidly controlling the cell designs. Power and ground rails occurred at the top and the bottom of cells in metal-2, and bit slices were arranged so one rail was shared between two adjacent slices. Data generally flowed horizontally through the cells in metal-1, while control signals and clock lines run vertically in polysilicon. The use of metal-2 was restricted within cells to the rails, so that it could be freely used for horizontal interconnect running over the cells without any design rule violations.

Cell height was selected based on the example of past work, and by building sample cells using different heights. An optimal value of 86λ was selected. Each cell was to be self-sufficient, so all required well/substrate contacts were internal. Port locations and boundary clearances were standardized, and multiple instances of ports was encouraged to ease cell composition.

All library cells were defined and exhaustively simulated in an implementation of Bryant's switch level simulator "Mossim" [6]. Layout cells used the same root name for ports as the Mossim definition, and a one-to-one correspondence between the two definitions was attempted through all levels in the design hierarchy. The XOR cell was an exception, since Mossim could not correctly model the 6 transistor design actually used.

Every signal that was considered reasonably useful for testing was routed through the shift registers. This included all read, write (except the write memory signal which was initially expected to be exported directly), and alu signals, the status flags from the datapath, and additionally, the mpc contents (the lines between the **mpc** register and the ROM), and the control signals for the 4×1 MUX feeding the **nextmpc** register, and controlling the microinstruction stack. In total, 72 bits are trapped by the shift registers. The only value passing between these components that was not trapped was the machine instruction code. Pass through lines were provided in the shift registers for this signal. Additionally, the control signal for the bidirectional pads, which was added at a late stage in the design, was not trapped. This omission made examining datapath register contents difficult, as the values could not be easily driven off chip, so the *wmem* signal was also routed through the shift registers in a later version.

The original die size and package had a limit of 64 pins. Bidirectional pins (32) were used for the bus to memory link. The MAR output required 14 pins, system clocks 2 pins, control inputs 2 pins, state output signals 2 pins, shift register controls, input, and output a total of 4 pins. 2 pins were used for the separate shift register clocks, instead of using one pin for a clock control input. Lastly, 2 power and 2 ground pins were used. A pair of power and ground pins (called *dirty power* and *dirty ground*) drive the pads only, to reduce noise on the supply lines to the chip, while the other pair drives the rest of the chip. In the final layout, the distribution of the pads around the chip perimeter was constrained by the number of bonding fingers along each cavity edge of the package, and a maximum 45 degree angle of bonding wires.

Simple input and output pads contain no logic, aside from buffering outputs, but the design of the bidirectional pads required more effort. When used in output mode, the pad had to increase the drive strength of the signal. A step-up buffer was used for this, but it must not drive in input mode. Thus the circuitry turns off both n and p-transistors by providing 0 and 5 volt gate inputs respectively. The designs were simulated using SPICE, and switching times in the 20 nanosecond range were achieved using a load capacitance of 50 pf. This speed was quite acceptable, given the constraints stated earlier.

3.6   Summary and Status

We have described the implementation in enough detail to let the reader appreciate the range of issues encountered in the design process, and to demonstrate that the design can be derived, in conjunc-

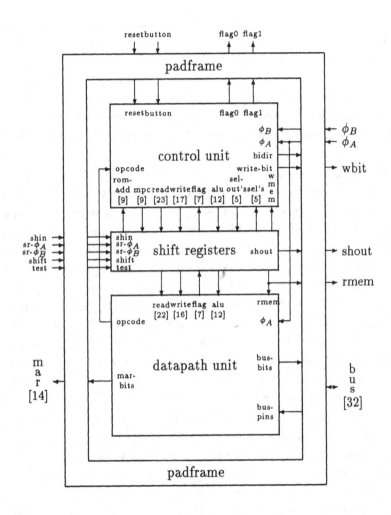

Figure 3.3: SECD Chip Major Subcomponents

tion with basic choices of architecture, by a fairly simple transformation process. The emphasis on the higher level view of the design, with a much less detailed look at layout, is in agreement with the focus of the verification on the top and register transfer level views of the design. The detailed description of the design given will aid the reader in understanding the formal definition of the system presented in the following chapter. Of particular importance is the timing scheme, and the relation between the levels which are formally described.

The scope of the project described was considerable, and involved a team of up to 9 individuals at various times. Two versions of the SECD chip have been fabricated. The first had a design error in the shift register component which made the chip entirely nonfunctional. This was purely a layout error, where the layout did not correspond to the Mossim model. The second design also has a flaw in a primitive gate, and once again, the cell layout did not correspond to the Mossim description of the circuit. At the time of writing, a revised layout has been checked, and extensive testing of the defective version of the chip, made possible by the separation of the major components, has been completed, indicating that all other components are fully functional.

Chapter 4

Formal Specification of the SECD Design

A microprocessor ultimately defines the operational semantics of a machine language. Specifying its behaviour therefore concerns the effect of executing machine instructions. However, just as we use many different levels of description in designing a device, we may and do use many different levels of specification. The formal specifications are in many ways rigorous versions of the informal descriptions. At the highest level, we want the specification to capture the designer's intent. At a lower level, we capture the structure of the design as an assembly of behaviours of simple components.

Microprocessors have been favoured subjects for the study of formal methods because they exhibit complex behaviours in both data and control operations. The SECD example differs most from previous subjects in the functional nature of the machine. Its support for procedures and recursion provide more complex instruction and memory state transitions than do traditional architectures. The chip is also significantly larger than many previous examples. For these reasons, management of complexity is a correspondingly larger issue.

The formal specification of the SECD chip consists of three levels of description:

- the **top level** specification, which defines the programming level model,

- the **register transfer level** definition, which defines the internal architecture in terms of multi-bit registers and functional (*i.e.*combinational logic) components, and

- the **low level** definition, consisting of primitive logic gates, single bit latches, and transistor networks for regular structures such as ROM's and PLA's.

The lowest level corresponds closely to the layout (and Mossim simulation) hierarchies, although the definitions of simple logic gates do not go all the way down to the transistor level. This is the model we wish to relate to the top level specification, but the problem size demands at least one intermediate level to abstract away some of the complexity. The intermediate level chosen corresponds very closely to the RTL view of the design development as pictured in Figure 3.2, and is the highest level which still expresses the internal architecture structure and control. Furthermore, the intermediate level chosen consists of an assembly of primitive components suitable for inclusion in a basic cell library. The lowest level definition matches the intermediate level in structure almost entirely, replacing component behaviour specifications with their more primitive implementations. Relating these two levels is largely concerned with independent component verification.

Because we wish to focus on control operation, this chapter concentrates on the RTL and top levels, limiting discussion of the low level to a few key elements that impact the sequel. We begin with a description of how hardware is modelled in HOL, and the data types and primitive operations common to all levels.

The complexity of the SECD chip makes it impossible within the scope of this book to give more than an outline of most of its component and data operation specifications: they are simply too large to be included in their entirety. The definitions alone comprise some 70 pages (4000 lines) of HOL source. All we can do is give a flavour of the work, and hope that enough information has been included to be meaningful to the reader.

A more thorough examination of this work requires examining the full HOL code. All the source files for the specification and verification of the SECD system described in this book (compatible with HOL88 Version 2.0) are available (at the time of writing) through the Internet network by anonymous ftp from the following host:

                              ftp.cl.cam.ac.uk

in the directory *hol*. It is also being included in the *contrib* directory of the HOL88 system distribution, beginning with version 2.1, which is also available via anonymous ftp from the same host.

4.1   Modelling Hardware

This research is concerned with bistate logic and logic devices, and uses *boolean* values to represent levels of electrical charge, with T and F representing 5v and 0v respectively. Signals that vary over time are represented by functions from discrete time, represented as :num values, to values of the appropriate type. For example, a single bit signal will be of type :num->bool. More complex data types can be built from single bits, particularly bit vectors. At a low level, we may describe a group of signals from *time* (*i.e.*:num) to :bool, while at a higher level we may prefer to describe a signal from *time* to a group of :bool values, and relate the two with an abstraction function.

We represent primitive combinational hardware devices in HOL by predicates which express logical relations on time-dependent input and output signals. For example, a 2 input NAND gate would be expressed by the relation:

NAND_spec a b c = !t:num. c t = ~(a t ∧ b t)

This expresses the behaviour that the output c at time t will be the value expressed by the *nand* relation on a and b at that same time. Universal quantification over the explicit time parameter indicates the relation holds at all points of time. Relations expressing devices with a memory component have an internal state signal, and relate signals at two different points of discrete time. For example, a simple *D-type* latch can be expressed by the relation: (where out is the state signal)

D-type clk in out =
!t:num. out(t+1) = clk t => in t | out t.

The composition of devices is expressed by conjunction with connected ports sharing the same parameter. Hidden nodes are expressed using existential quantification. For example, a 4 nand gate implementation of an *exclusive-or* function may be defined by the circuit:

```
XOR_imp a b c =
? n1 n2 n3:num->bool.
   (NAND_spec a  b  n1) ∧ (NAND_spec a  n1 n2) ∧
   (NAND_spec n1 b  n3) ∧ (NAND_spec n2 n3 c )
```

Extensive use is made of signals representing fixed numbers of bits, referred to as $word_n$ types.[1] This type is defined in terms of specified-width *buses* of discrete signals, where objects of type :(*)bus are defined using the constructors Wire for the base case (discrete signal), and Bus for the addition of each subsequent signal. They resemble nonempty lists. A specified $word_n$ type object, such as a :word14 value used to represent an address, is created by applying the constant Word14 to a :(bool)bus of *Width* 14. Word14 is an abstraction function from the *representing* type :(bool)bus. The corresponding *representation* function :Bits14 maps objects of type :word14 back to :(bool)bus type objects. Constants of $word_n$ type may be written as binary strings: #00000000000011 represents a 14 bit word that one may choose to interpret as the number 3.

A simple memory can be defined as a function having the type :word14->word32. Since new values can be written to memory, it is defined as a function from discrete time, :num->word14->word32. Other examples of nonprimitive data types include tuples and lists. Gordon [21] gives a general description of describing hardware in the HOL system.

As the unit of discrete time is different for each level of definition, we adopt a convention of subscripting the explicit time parameters to indicate the granularity of time intended, low level (fine grain) time t_f, RTL (medium grain) time t_m, and top level (coarse grain) time t_c.

4.2   The Top Level Specification

At the most abstract level, the SECD machine is defined in terms of transformations to S-expressions in the 4 stacks, as shown in Table 2.3. A formal specification of the top level behaviour is ideally defined in terms of transformations to an S-expression data type that closely resembles this elegant definition. The closer the resemblance the better we are assured that the HOL specification captures the intent.

---

[1]This data type definition was implemented by T. Melham, and differs from later implementations in not permitting "empty" word types.

4.2.1 A high level data type

The method used by SECD for implementing recursive function definitions as closures with a circular environment component raises the complexity of the data representation problem considerably. Such circular S-expression lists, created by a destructive operation, cannot be mapped to a simple recursive data type. Further, structure is shared by S-expressions, particularly the environment component of closures. Each mutually recursive function closure references the same environment, which is also in the E stack. When a destructive replace operation is performed (by executing a RAP instruction) to create the circular list structure, the change affects the common component of all closures simultaneously.

Thus, a much more primitive representation is used to describe the top level specification, following on the work of Mason [40, 41] on the semantics of destructive Lisp. Rather than directly defining transformations on structures of an S-expression data type, an abstract memory type is defined which can embed representations of S-expression data structures. Further, a set of primitive operations upon the memory is defined, corresponding to the operations on S-expressions, namely *cons, car, cdr, atom, rplaca, eq, leq, add*, and *sub*. The four state registers contain pointers to the appropriate S-expression representation. Finally, an additional *free* pointer to the free list structure is needed to define the *cons* operation.[2] The state of the machine is then defined by a tuple:

$$(S, E, C, D, Free, memory, FSM\_state).$$

where the *FSM_state* is one of the 4 major states of the top level finite state machine view of the machine (Figure 3.1).

The abstract memory type μ is basically a function type:

$$\mu = \delta \to (\delta \times \delta \cup \alpha)$$

where δ is the domain of the function and α is the set of atoms which will include the sets of *integers* and of *symbols*. The set of symbols

---

[2]This differs from Mason's semantics, wherein a *cons* operation enlarged the domain of a memory, and there is no explicit notion of a free list. See page 162 for some reflections on alternate definitions of data types.

Operation	Type
primitive operations	
M_Car, M_Cdr	$(\delta \times \mu \times \delta) \rightarrow (\delta)$
M_Cons	$(\delta \times \delta \times \mu \times \delta) \rightarrow (\delta \times \mu \times \delta)$
M_Eq, M_Leq	$(\delta \times \delta \times \mu \times \delta) \rightarrow bool$
M_Add, M_Sub	$(\delta \times \delta \times \mu \times \delta) \rightarrow (\delta \times \mu \times \delta)$
M_Replaca, M_Replacd	$(\delta \times \delta \times \mu \times \delta) \rightarrow (\delta \times \mu \times \delta)$
M_setm, M_setf	$(bool \times \delta \times \mu \times \delta) \rightarrow (\delta \times \mu \times \delta)$
extractor functions	
M_mark, M_field	$\delta \rightarrow \mu \rightarrow bool$
M_Int_of	$(\delta \times \mu \times \delta) \rightarrow integer$
M_Atom_of	$(\delta \times \mu \times \delta) \rightarrow \alpha$
predicates	
M_Atom	$(\delta \times \mu \times \delta) \rightarrow bool$
M_is_cons	$(\delta \times \mu \times \delta) \rightarrow bool$
M_is_T	$(\delta \times \mu \times \delta) \rightarrow bool$
auxiliary functions	
M_garbage_collect	$(\mu \times \delta) \rightarrow (\mu \times \delta)$
M_CAR, M_CDR	$(\delta \times \mu \times \delta) \rightarrow (\delta \times \mu \times \delta)$
M_Cons_tr	$(\delta \times \delta \times \mu \times \delta) \rightarrow (\delta \times \mu \times \delta)$

Table 4.1: Primitive Operations on Abstract Memories[3]

includes the symbolic constants: T, F, and NIL. The definition of
memory is extended to incorporate garbage collection features by
adding *mark* and *field* bits to each cell:

$$\mu = \delta \rightarrow ((bool \times bool) \times (\delta \times \delta \cup \alpha)).$$

Additionally, the garbage collection operations *replacd*, *setf*, and *setm*
are included, as well as a *garbage_collect* function, which is left undefined for this proof effort. Extractor functions for the *mark* and
field bits, and the *integer* and *atom* fields are provided for the values
returned by the μ function.

The type α is represented in HOL by the type :atom, defined by
the simple grammar:

```
atom = Int integer | Symb num,
```

---

[3]For convenience of notation, we use the abbreviations μ and δ to represent
the HOL data types :(word14,atom)mfsexp_mem and :word14 respectively.

where Int and Symb are type constructors. The three symbolic constants are defined as follows.

```
NIL_atom = Symb 0        T_atom = Symb 1        F_atom = Symb 2
```

In order to represent the type μ in HOL, we define an abstract polymorphic memory type : (*,**)mfsexp_mem by mapping it to an existing type :*->((bool#bool)#(*#* + **)). The specific type required is the instance : (word14,atom)mfsexp_mem, the domain of which matches that of the lower level representation of memories. Both defined types have associated REP and ABS functions to map between the abstract and representing types. For example, the term REP_mfsexp_mem(memory:(word14,atom)mfsexp_mem) has type:

   :word14->((bool#bool)#(word14#word14 + atom)).

The relevant built-in functions on abstract memories and their types are summarised in Table 4.1. To distinguish these abstract memory functions, the function names uniformly begin with "M_". For consistency, all functions take both a memory (μ) and the free list pointer (δ) as the last items in the tuple argument, although the free pointer is not always used. Operations such as M_Car, M_Cdr, M_Eq, M_Leq, etc. do not alter memory, while M_Cons, M_Add, M_Sub, M_setm, M_setf, M_Replaca and M_Replacd do alter one cell in the memory, and thus must return the new memory.

```
M_Cons (x,y,MEM,free) =
  let (Mem,Free) = (M_Atom (free,MEM,free)
                 => (M_garbage_collect (MEM, free))
                  | (MEM, free))
  in ( Free,
       (@m:(*,**)mfsexp_mem.
         REP_mfsexp_mem m =
         (λa. (a=Free) => (F,F),INL(x,y)
                       | REP_mfsexp_mem Mem a)),
      M_Cdr (Free, Mem, Free))

M_Car (x,MEM,free) = FST (OUTL (SND (REP_mfsexp_mem MEM x)))
```

In the definition of M_Cons above, observe the use of the select operator "@" to specify a memory of the abstract :mfsexp_mem data type in terms of the operation on the corresponding object in the representative data type (obtained by applying REP_mfsexp_mem). The INL function is the left injector for sum datatypes. Similarly, in defining M_Car the memory object is transformed to the representative data type, applied to a cell (:word14) returning an object of type :(bool#bool)#(word14#word14+atom), selects the second part of the pair and extracts the left summand type (:word14#word14), and takes the first part of this pair as the car (the second part is correspondingly the cdr).

```
M_Eq (x,y,MEM,free)  =
   let Atom_x = M_Atom(x,MEM,free)
   and Atom_y = M_Atom(y,MEM,free)
   in ( (Atom_x ∧ ¬Atom_y)
        => F
        | (¬Atom_x ∧ Atom_y)
           => F
           | (M_atom_of(x,MEM,free) = M_atom_of(y,MEM,free)))
```

The M_Eq function is partial: at least one argument must be atomic. This precisely matches the semantics of Henderson's original definition [28], and corrects the imprecise informal treatment on page 17.

```
M_Add (x,y,MEM,free) =
   let x_val = M_int_of (x,MEM,free)
   in
   let y_val = M_int_of (y,MEM,free)
   in
   M_store_int (x_val modulo_28_Add y_val, MEM, free)
```

The M_Add function first extracts two records from the memory, expecting them to be integer constants. The arithmetic implemented in the design does not record overflow or underflow, performing instead arithmetic on a finite cyclic group, with the integers from -2^{27} to $2^{27} - 1$, thus accounting for the infix modulo_28_Add function in place of the ordinary integer plus operator. (See page 80 for definitions of the modulo operations.) The function **M_store_int** is very similar to the **M_Cons** function, except it installs an integer atom record in memory instead of a cons record.

```
M_Rplaca (c,v,MEM,free) =
  ( c,
    (@m:(*,**)mfsexp_mem. REP_mfsexp_mem m =
       (λa. (a = c) => (F,F),INL(v, M_Cdr(a, MEM, free))
                     | REP_mfsexp_mem MEM a)),
    free)
```

The destructive replace operation **M_Rplaca** substitutes the supplied argument **v** for the *car* field of the record **c** by overwriting. The specification closely resembles that of the **M_Cons** function.

A few additional functions are given in Table 4.1. The **M_CAR** and **M_CDR** functions permit composition of the primitive memory operations, returning the unaltered memory and free pointer. For example, to access an argument to a **LD** command, we can write:

let m = M_Int_of(M_CAR(M_CAR(M_CDR(c,MEM,free)))).

The function **M_Cons_tr** transposes the first 2 arguments of the **M_Cons** function, to permit composition where the memory and free pointer arguments are handed on from the computation of the first *cons* argument instead of the second.

4.2.2 Instruction semantics

Using the data types and operators just defined, we may now proceed to define the desired semantics for each of the machine instructions. We begin with the **LDF** instruction in figure 4.1. The functions **cell_of, mem_of, free_of** (and also **mem_free_of**) are extractor functions from the triples returned by most **M_** functions. Compare this definition with the original specification in Table 2.3.

The **LD** instruction semantics (Figure 4.2) require a look-up in the environment at a location specified by in-line arguments to the instruction. The local variables **m** and **n** are bound to the arguments, after first extracting integers from the word representations, and then casting the result as a type **:num** value. This data transformation is required because the iterative function **nth**, used to specify repeated

```
LDF_trans (s:δ,e:δ,c:δ,d:δ,free:δ,MEM:μ) =
 let cell_mem_free_1 = M_Cons_tr(e,M_CAR(M_CDR(c,MEM,free)))
 in
 let cell_mem_free_2 = M_Cons_tr(s,cell_mem_free_1)
 in
 (cell_of cell_mem_free_2,                         % s %
  e,                                               % e %
  M_Cdr(M_CDR(c,mem_free_of cell_mem_free_2)),     % c %
  d,                                               % d %
  free_of cell_mem_free_2,                         % free %
  mem_of cell_mem_free_2,                          % memory %
  top_of_cycle)                                    % state %
```

Figure 4.1: Transition for **LDF** Instruction

```
LD_trans (s:δ,e:δ,c:δ,d:δ,free:δ,MEM:μ) =
 let m = pos_num_of(M_int_of(M_CAR(M_CAR(M_CDR(c,MEM,free)))))
 in
 let n = pos_num_of(M_int_of(M_CDR(M_CAR(M_CDR(c,MEM,free)))))
 in
 let cell_mem_free =              % returns (cell,(MEM,free)) %
     M_Cons_tr (s, M_CAR (nth n M_CDR
                     (M_CAR (nth m M_CDR
                             (e, MEM,free)))))
 in
 (cell_of cell_mem_free,,                          % s %
  e,                                               % e %
  M_Cdr (M_CDR (c, mem_free_of cell_mem_free)),    % c %
  d,                                               % d %
  free_of cell_mem_free,,                          % free %
  mem_of cell_mem_free,,                           % memory %
  top_of_cycle)                                    % state %
```

Figure 4.2: Transition for LD Instruction

applications of M_CDR, is defined by primitive recursion on the type
:num. The function **pos_num_of** is a partially defined (but total)

```
RAP_trans (s:δ,e:δ,c:δ,d:δ,free:δ,MEM:μ) =
  let cell1_mem_free = M_Cons_tr(d,M_CDR(c,MEM,free))
  in
  let cell2_mem_free = M_Cons_tr(cell_of cell1_mem_free,
                                 M_CDR(e,mem_free_of cell1_mem_free))
  in
  let d_mem_free = M_Cons_tr(cell_of cell2_mem_free,
                             M_CDR(M_CDR(s,mem_free_of cell2_mem_free)))
  in
  let e_mem_free = M_Rplaca(M_Cdr(M_CAR(s,mem_free_of d_mem_free)),
                            M_CAR(M_CDR(s,mem_free_of d_mem_free)))
  in
  (NIL_addr,                                     % s %
   cell_of e_mem_free,                           % e %
   M_Car(M_CAR(s,mem_free_of e_mem_free)),       % c %
   cell_of d_mem_free,                           % d %
   free_of e_mem_free,                           % free %
   mem_of  e_mem_free,                           % memory %
   top_of_cycle)                                 % state %
```

Figure 4.3: Transition for **RAP** Instruction

function: it may only be evaluated when applied to nonnegative integers. Such a "loose" specification avoids the possibility of errors going unnoticed when arbitrary values are returned from what are considered "don't care" cases that mistakenly arise. If for example **pos_num_of** is inadvertently applied to a negative integer, we would like to know about it. By preventing its evaluation, we shall.

One of the most complex transitions is the **RAP** instruction (Figure 4.3). Four memory locations are changed, three of them are the result of cons operations, and the fourth is the destructive replace operation used to create the circular environment data structure for mutually recursive function definitions.

The **STOP** instruction specification (Figure 4.4) is the only one remaining which differs in form from those just given. Instead of a *function* from one state to another, it is a *relation* between two states. In the first attempt to complete the verification, it was discovered

```
STOP_trans_relation ((s':δ,e':δ,c':δ,d':δ,free':δ,MEM':μ,state'),
                     (s:δ,e:δ,c:δ,d:δ,free:δ,MEM:μ)) =
  (s' = M_Car(s,MEM,free)) ∧
  (e' = e) ∧
  (c' = c) ∧
  (d' = d) ∧
  (free' = free) ∧
  (!a. (a = NUM_addr)
       => (!z. M_is_cons(a,MEM',z) ==> (M_Cdr(a,MEM',z) = s'))
       | (REP_mfsexp_mem MEM' a = REP_mfsexp_mem MEM a)) ∧
  (state' = idle)
```

Figure 4.4: Transition for STOP Instruction

that the functional specification was not provably implemented, because the machine returns a pointer to the result of the computation in the highest memory address, by writing only the *cdr* field of the word. The other lines of the bus are not driven during this operation. As a result, the value in that memory location is only partially determined, and the transition cannot be specified in the same form as the other instructions.

This feedback from the verification to the specification is not necessarily a bad thing. Indeed, the only problem was that the initial specification was too strict. Further comments on the general nature of specification are left to the concluding chapter.

4.2.3   The top level definition

The top level specification In Figure 4.5 is written in two parts: the next state of the machine is defined for each instruction, using the set of 21 semantic definitions, and the top level system specification is defined, closely resembling the top level finite state machine of Figure 3.1, including an assurance of starting out in the *idle* state. In both cases, these are defined as relations. The types of all parameters have the subscript "$_c$", identifying them as signals at a *coarse* grain of time.

```
NEXT((s':δc,e':δc,c':δc,d':δc,free':δc,MEM':μc,state':statec),
     (s:δc,e:δc,c:δc,d:δc,free:δc,MEM:μc)) =
 let instr = M_int_of(M_CAR(c,MEM,free))
 in ((instr = LD)   =>
     (s',e',c',d',free',MEM',state' = LD_trans(s,e,c,d,free,MEM))
     ...
    |(instr = LEQ)  =>
     (s',e',c',d',free',MEM',state' = LEQ_trans(s,e,c,d,free,MEM))
    |(STOP_trans_relation
        (s',e',c',d',free',MEM',state'),(s,e,c,d,free,MEM)))
```

```
SYS_spec (s:δc) (e:δc) (c:δc) (d:δc) (free:δc)
        (MEM:μc) (button_pin:boolc) (state:statec) =
 (state 0c = idle) ∧
 (!tc.
 (λ(s',e',c',d',free',MEM',state').
   (state tc = idle)
   => (button_pin tc
      => (s',e',c',d',free',MEM',state' =
          M_Cdr(M_CAR(NUM_addr,MEM tc,free tc)),NIL_addr,
          M_Car(M_CAR(NUM_addr,MEM tc,free tc)),NIL_addr,
          M_Cdr(NUM_addr,MEM tc,free tc),MEM tc,top_of_cycle)
       | (s',e',c',d',free',MEM',state' =
          s tc,e tc,c tc,d tc,free tc,MEM tc,idle))
   |(state tc = error0)
    => (button_pin tc
       => (s',e',c',d',free',MEM',state' =
           s tc,e tc,c tc,d tc,free tc,MEM tc,error1)
        | (s',e',c',d',free',MEM',state' =
           s tc,e tc,c tc,d tc,free tc,MEM tc,error0))
   |(state tc = error1)
    => (button_pin tc
       => (s',e',c',d',free',MEM',state' =
           s tc,e tc,c tc,d tc,free tc,MEM tc,error1)
        | (s',e',c',d',free',MEM',state' =
           s tc,e tc,c tc,d tc,free tc,MEM tc,idle))
   | NEXT ((s',e',c',d',free',MEM',state'),
           (s tc,e tc,c tc,d tc,free tc,MEM tc)))
 (s(tc+1),e(tc+1),c(tc+1),d(tc+1),free(tc+1),MEM(tc+1),state(tc+1))))
```

Figure 4.5: Top Level Specification

4.3   The Low Level Definition

The low level definition uses simple gates as base components, typically with functions of the complexity of AND, OR, XOR, etc., and bears a one-to-one relation to the layout of gates in the SECD chip.

The representation of time uses the coarsest granularity that captures the behaviour of the clock phases: using 4 points per clock cycle, one for each phase and one point taken between each pair of phases.

$$\phi_A \qquad \phi_B$$

The presence of two separate clocks, one for the normal mode of chip operation, and the other operating the shift registers for test mode, confuses the representation. The two clocks should never cycle simultaneously, and by constraining their behaviour appropriately, the atomic interval at this granularity relates to a cycling of either clock. An informal description of these constraints follows.

- A clock cycle always consists of nonoverlapping assertions of ϕ_A followed by ϕ_B, for either clock.

- The separate clocks do not cycle simultaneously. Once started each clock always completes its cycle uninterrupted by the other clock. Typical behaviour resulting from these two constraints is illustrated by the following timing diagram.

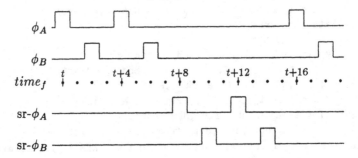

- The single exception to the first item is that the system clock phase ϕ_B is asserted at power-up. The reset input is asserted at this time as well, forcing the chip into a known initial state (see Figure 4.8).

- Lastly, since the verification will only concern normal mode of operation, the system clock will be the only clock that cycles.

Although restricting the chip to normal mode of operation, these constraints made it clear that intervals of the fine granularity of time correspond to advancing *either* clock.

Formalising this description is straightforward, first defining a set of predicates to describe the clock phase line behaviours.

```
! t_f (f:num->bool).
  CycleA t_f f = f t_f ∧ ¬f(t_f+1) ∧ ¬f(t_f+2) ∧ ¬f(t_f+3)

! t_f (f:num->bool).
  CycleB t_f f = ¬f t_f ∧ ¬f(t_f+1) ∧ f(t_f+2) ∧ ¬f(t_f+3)

! t_f (f:num->bool).
  noCycle t_f f = ¬f t_f ∧ ¬f(t_f+1) ∧ ¬f(t_f+2) ∧ ¬f(t_f+3)

! phi_A phi_B t_f.
  CYCLE phi_A phi_B t_f = (CycleA t_f phi_A) ∧ (CycleB t_f phi_B)

! phi_A phi_B t_f.
  no_CYCLE phi_A phi_B t_f =
  (noCycle t_f phi_A) ∧ (noCycle t_f phi_B)
```

The constraint for general operation of the chip clock lines is completed by determining the behaviour at times 0 and 1, and then defining it for each subsequent interval.

```
! phi_A phi_B sr_phi_A sr_phi_B.
  proper_clocking phi_A phi_B sr_phi_A sr_phi_B =
  (¬phi_A 0 ∧  phi_B 0 ∧ ¬sr_phi_A 0 ∧ ¬sr_phi_B 0) ∧
  (¬phi_A 1 ∧ ¬phi_B 1 ∧ ¬sr_phi_A 1 ∧ ¬sr_phi_B 1) ∧
  (!t_m n.
    let t_f = 2 + (n * t_m)
    in
    (   CYCLE phi_A phi_B t_f ∧ no_CYCLE sr_phi_A sr_phi_B t_f ∨
     no_CYCLE phi_A phi_B t_f ∧    CYCLE sr_phi_A sr_phi_B t_f))
```

Restricting this to normal mode of operation requires only deleting
the last disjunct. Notice that the quantified variable t_m can be
viewed as a medium grain time index, with the time t_f representing
the corresponding point in fine grain time.

The behaviour of level triggered latches at this time granularity
would be captured adequately by the D-type definition given in section 4.1, with a clock phase line as its first argument. However, the
standard design regimen for 2-phase non-overlapping clocking, that
of linking memory devices clocked on opposite clock phases with combinational logic, is violated in one part of the SECD chip design. Two
registers clocked on the same clock phase are wired in series, with
the output of the first feeding the input of the second through combinational logic. This odd circuit feature degrades chip performance,
but at sufficiently slow clock rates the design can still function properly since the output of a level triggered latch changes at the start
of the clock pulse and no circular feedback path exists. This single
piece of questionable design affects the entire chip specification and
adds an obligation to prove the impossibility of circular feedback. A
modified definition of the primitive latch is required to express this
behaviour appropriately, expressing the present state as a function
of the clock signal and input at the present moment instead of the
previous moment.

```
latch (clk:num->bool) (inp:num->bool) (state:num->bool) =
  !t_f. state t_f = (clk t_f) => inp t_f | state (PRE t_f )
```

The regular structures, such as the microcode ROM, decoders,
and a PLA can be defined as logical functions, but their complexity
is quite different from the rest of the primitive gates. These regular
structures are produced by automated layout tools, and the transistor network produced is specified in HOL. The use of a pseudo
nMOS OR plane in the PLA and ROM requires the use of a multi
level logic, to capture the floating value which will "pull up" to 5v in
the actual device. The output of the device will always be at a fully
restored level, so the interface converts back to a boolean data type.

4.4 Register Transfer Level Specification

The Register Transfer Level (RTL) model of the chip is similar in component hierarchy to the low level view, but does not have the same depth. The lowest items in the RTL hierarchy include devices such as multi-bit registers, which are nonprimitive components of the low level hierarchy. The chip has three major components: the control unit, the datapath, and the pad frame (CU, DP, and PF respectively). Generally, the primitive components are behavioural specifications for the corresponding components at the low level.

Notably absent from this view is the shift register component. Under normal operation, these are intended to be transparent to the operation of the chip, and thus constraining their controls and the clocks appropriately at the lower level permits abstracting them out of this level of description completely. However, for future compatibility with an extended specification that can include debug mode, a clock signal SYS_Clocked is included to represent the cycling of the system clock lines. This signal parameterises every memory device in this level view.

4.4.1 Temporal representation

The time granularity at this level corresponds to clock cycles. Selecting which point of the cycle to use for this time grain is complicated by the outputs of registers clocked on different clock phases being multiplexed into the same register input (*i.e.* this is the case for the MPC9 register). The part of the clock cycle over which the output of the register is stable is determined by the phase clocking the register. In order to have a uniform temporal abstraction for the time parameter for each input type, we select the points in the finer grain when the ϕ_A phase is asserted to represent the latch state at the RTL (medium) time granularity. We will come back to the reasons after describing the two types of registers.

Two different types of registers are used. The control unit uses pairs of latches clocked on opposite clock phases, while the datapath registers clock write operations on ϕ_A to prevent transitory spikes on state transitions from causing unwanted writes. The low level circuit

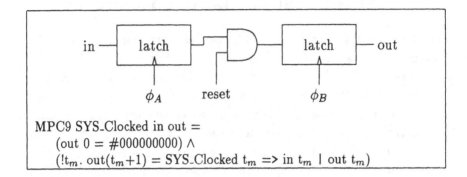

Figure 4.6: Control Unit Register Schematic and Definition

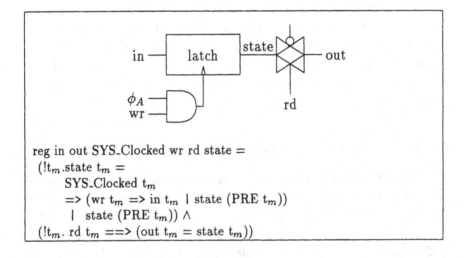

Figure 4.7: Datapath Register Schematic and Definition

and the RTL specification for each are shown in Figures 4.6 and 4.7.

In both definitions, the clock signal SYS_Clocked is an abstraction of the two distinct clock phases (ϕ_A and ϕ_B), and is asserted when the system clock cycles. (Under the lower level clock constraints, the system clock always cycles.) Note that the MPC9 register specification does not include the *reset* input explicitly. However, the lower level constraints on the *reset* signal force an initialisation at power-up, and this appears as the determined value #000000000 in the register at time $t_m = 0$. The typical datapath register has a gated output, and is clocked by both the system clock phase and the specific write signal. The clock signal is included here to permit eventual extension of the verification to the test mode of chip operation. The difference in the definition of the registers and the choice of point on the time cycle can be seen from the timing diagram in Figure 4.8.

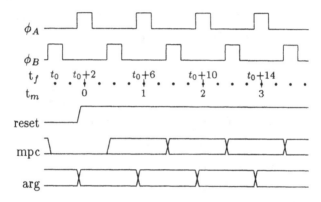

Figure 4.8: Relating low and RTL times

The *reset* signal is held low at the start of the clock operation. The MPC9 register content (*i.e.* mpc) at $t_m + 1$ is a function of the inputs at time t_m. The ARG register (a typical datapath register) is used to pass the instruction opcode read from memory to the MPC9 input, all during the same pulse of ϕ_A. Another input to MPC9 is the increment of its current value *mpc*. Consider both possible inputs at $t_m = 3$. The value had to be supplied at the input of the first of the pair of latches at fine grain time $t_0 + 10$, and is passed to the second (slave) latch at time $t_0 + 12$. Whether the input is supplied by writing through the ARG register as it is being read from memory, or it is supplied by incrementing the value held in the second latch, we get the relation mpc 3 = opcode(arg 2) or mpc 3 = Inc9(mpc 2) at the RT level, just as desired. Sampling one (fine grain) point earlier would have required distinct temporal abstractions for signals arising from registers clocked on different clock phases.

4.4.2 The datapath specification

The definition of the datapath begins with itemizing the data types
introduced, specified constants, and definitions of primitive operations. Component definitions and the composition of the whole follow. Refer to Figure 3.2 for a visual representation of all the datapath
components and their organization.

Four **fixed length word types** are needed for the datapath,
:word32 for representing words from memory, :word2 for record type
and garbage collection bits fields of words, :word14 for memory addresses, and :word28 for atom value fields. Several *word$_n$* constants
are defined below.

```
T_addr    = #00000000000001       RT_SYMBOL = #10
F_addr    = #00000000000010       RT_NUMBER = #11
NIL_addr  = #00000000000000       RT_CONS   = #00
NUM_addr  = #11111111111111
ZEROS14   = #00000000000000
ZERO28    = #0000000000000000000000000000
```

The first 4 are the addresses of reserved words in memory. ZEROS14
is used to pad a 14 bit address to a 28 bit number. ZEROS28 is
the representation of the integer 0, and the remaining three are the
record type identifiers.

Field extractor functions extract fields from larger words.

```
car_bits (Word32(Bus b32 (Bus b31 (Bus b30 (Bus b29 (Bus b28
                  (Bus b27 (Bus b26 (Bus b25 (Bus b24 (Bus b23
                  (Bus b22 (Bus b21 (Bus b20 (Bus b19 (Bus b18
                  (Bus b17 (Bus b16 (Bus b15 (Bus b14 (Bus b13
                  (Bus b12 (Bus b11 (Bus b10 (Bus b9  (Bus b8
                  (Bus b7  (Bus b6  (Bus b5  (Bus b4  (Bus b3
                  (Bus b2  (Wire (b1:bool)
                  )))))))))))))))))))))))))))))))) =
Word14(Bus b28 (Bus b27 (Bus b26 (Bus b25 (Bus b24
                  (Bus b23 (Bus b22 (Bus b21 (Bus b20 (Bus b19
                  (Bus b18 (Bus b17 (Bus b16 (Wire b15)))))))))))))))
```

The *word$_n$* types offer straightforward extractor definitions, without
the need for obscure conversions to different data types typical of an

earlier *word_n* implementation, as shown by the above definition of
car_bits. Other functions are cdr_bits, atom_bits, garbage_bits,
mark_bit, field_bit and rec_type_bits.

Recognizer functions compare the record type field of a word
with the record type constants: is_symbol, is_number, is_cons,
and is_atom.

```
is_symbol (x:word32) = (rec_type_bits x = RT_SYMBOL)

is_number (x:word32) = (rec_type_bits x = RT_NUMBER)

is_cons (x:word32) = (rec_type_bits x = RT_CONS)

is_atom (x:word32) = (is_symbol x ∨ is_number x)
```

Conversion functions to map bit values to numbers and integers
are defined, including bv, Val and iVal. The latter two of which are
defined using val, which uses an accumulating parameter.

```
bv (x:bool) = x => 1 | 0

(val (n:num) ((Wire w):(bool)bus)  =  2 * n + (bv w)) ∧
(val n        (Bus b bus)          =  val (2 * n + (bv b)) bus)

Val = val 0

(iVal ((Wire w):(bool)bus)    =  neg (INT (bv w))) ∧
(iVal (Bus b bus) =
  INT (val 0 bus)   minus INT ((2 EXP (Width bus)) * bv(b)))
```

Lastly, **constructor functions** for each record type are defined,
using the Concat function to join *buses* together.

```
bus32_cons_append (a:word2)(b:word2)(c:word14)(d:word14) =
Word32(Concat(Bits2 a)
             (Concat(Bits2 b)(Concat(Bits14 c)(Bits14 d))))

bus32_num_append (a:word28) =
Word32(Concat(Bits2 #00)(Concat(Bits2 RT_NUMBER)(Bits28 a)))
```

```
bus32_symb_append (a:word28) =
Word32(Concat(Bits2 #00)(Concat (Bits2 RT_SYMBOL)(Bits28 a)))

gc_bus32_append (a:bool)(b:bool)(c:word32) =
Word32 (Bus a (Bus b (Tl_bus (Tl_bus (Bits32 c)))))
```

The 4 constant "registers": NUM, Nil, TRUE, and FALSE, as well as the *clearunit* that pads a 14 bit address with zeros to 28 bits, are all implemented with busgates. A busgate is simply a set of transmission gates, controlled by a *read* signal. A polymorphic version is defined as follows, and the type variable instantiated as determined by its arguments.

```
busgate (in_val:num->*)(rd:num->bool)(out:num->*) =
!t_m. (rd t_m) ==> (out t_m = in_val t_m)
```

When used to gate a constant value onto the bus, the predicate is applied to an abstraction of the form $(\lambda t_m . \texttt{NIL\_addr})$. The READ_MEM component is also a busgate, which connects the bus to the output of the bidirectional i/o pad. This is required to isolate the pad's on-chip output from the bus when a read memory instruction is not asserted, since the pads default to input mode of operation.

The form of the register definition was described earlier. As with the busgate, a polymorphic register is defined, as well as a variant with input and output connected to the same node, which is typical for most registers connected to the bus. The definitions follow.

```
reg (in_sig:num->*)(out_sig:num->*)(clocked:num->bool)
    (wr:num->bool)(rd:num->bool)(st:num->*) =
(!t_m. st t_m = clocked t_m
                   => (wr t_m => in_sig t_m | (st (PRE t_m)))
                   | (st (PRE t_m))) /\
(!t_m. (rd t_m) ==> ((out_sig t_m) = (st t_m)))

register (sgl:num->*) = reg sgl sgl
```

Instances of these polymorphic registers are defined for each required register, instantiated with the appropriate type (for 2, 14, and 32 bit words), typically labelled S_reg, E_reg, CAR_reg, etc.

The **Cons** unit is simply implemented with 4 busgates: two 14 bit busgates connect the car and cdr fields inputs, while two 2 bit busgates are connected to constants for the garbage bits and record type bit fields. The output is a 32 bit value, with appropriate field selectors applied to select the connection for each busgate.

The **FLAGSUNIT** performs various tests on data values and returns 7 status flags for use by the control unit. Only the LEQ operation needs definition; the others test particular bits or compare all bits. **LEQ_prim** compares the integer values represented by the words being compared.

```
LEQ_prim (x:word32) (y:word32) =
  let ival_x = (iVal o Bits28 o atom_bits) x
  in
  let ival_y = (iVal o Bits28 o atom_bits) y
  in
  ((ival_x below ival_y) V (ival_x = ival_y))

FLAGSUNIT bus arg atomflag bit30flag
          bit31flag zeroflag nilflag eqflag leqflag =
  !t_m.
  (atomflag   t_m = is_atom (bus t_m))              ∧
  (bit30flag  t_m = field_bit (bus t_m))            ∧
  (bit31flag  t_m = mark_bit (bus t_m))             ∧
  (zeroflag   t_m = (atom_bits (bus t_m) = ZERO28) ) ∧
  (nilflag    t_m = (cdr_bits (bus t_m) = NIL_addr)) ∧
  (eqflag     t_m = (bus t_m = arg t_m))            ∧
  (leqflag    t_m = LEQ_prim (arg t_m) (bus t_m))
```

The **ALU** can perform 9 distinct operations. Six of these are used exclusively in garbage collection: *replcar, replcdr, setbit31, setbit30, resetbit31,* and *resetbit30*. The remaining operations are the arithmetic operations: *add, sub,* and *dec*. The *mul, div,* and *rem* operations are not implemented, and selecting any of these defaults to the *dec* operation.

All 6 of the garbage collection operations are destructive operations, replacing a bit or field of the operand word.

```
REPLCAR (x:word32) (y:word14) =
  bus32_cons_append (garbage_bits x)
        (rec_type_bits x) y (cdr_bits x)

REPLCDR (x:word32) (y:word14) =
  bus32_cons_append (garbage_bits x)
        (rec_type_bits x) (car_bits x) y

SETBIT30 (x:word32)   = gc_bus32_append (mark_bit x)  T x

SETBIT31 (x:word32)   = gc_bus32_append T (field_bit x) x

RESETBIT30 (x:word32) = gc_bus32_append (mark_bit x)  F x

RESETBIT31 (x:word32) = gc_bus32_append F (field_bit x) x
```

The SECD chip uses 2's complement (28 bits) representation for integer values. Overflow and underflow are not recognized, instead the arithmetic operations wrap by truncating the upper bits. In somewhat more formal mathematical terminology, it is a finite cyclic group, with the set of integers in the range (-2^{27}) through $(2^{27} - 1)$, and an addition operation defined in terms of addition on the integers, but mapped onto the group as follows (for additions performed on integers in the group):

- integers in the range (-2^{27}) through $(2^{27} - 1)$ are mapped to the same integer in the group,

- integers in the range (-2^{28}) through $(-2^{27} - 1)$ are mapped to a nonnegative integer in the group by adding 2^{28},

- and integers in the range 2^{27} through $(2^{28} - 2)$ are mapped to a negative integer in the group by subtracting 2^{28}.

The arithmetic operations at the top level of the specification are defined as normal integer operations, with the result mapped onto the integers within the group as described above. One slight generalisation accommodates all integer results by first performing a mod 2^{28} operation to the magnitude of the integer to bring it into the range for which the previous mapping applies. The required definitions follow.

```
normalize (n:num)(b:num) = b MOD (2 EXP n)

norm28 (i:integer) =
  (NEG i => let n' = normalize 28 (pos_num_of(neg i))
     in
               (n' <= (2 EXP 27) => neg(INT n')
                                  | INT((2 EXP 28) - n'))
          | let n' = normalize 28 (pos_num_of i)
            in
            (n' < (2 EXP 27) => INT n'
                              | neg(INT((2 EXP 28) - n')))))

(i1:integer) modulo_28_Sub (i2:integer) = norm28 (i1 minus i2)

(i1:integer) modulo_28_Add (i2:integer) = norm28 (i1 plus i2)

modulo_28_Dec (i1:integer) = norm28 (i1 minus (INT 1))
```

The detailed definition of the machine implementation of the arithmetic operations is confined to the lowest level only. At the RT level, the arithmetic operations on low level data types (:**word28**) are defined in terms of the corresponding integer operation on the integer representation of the arguments.

```
ADD28 (x:word28)(y:word28) =
  (let ival_x = (iVal o Bits28) x
   in
   let ival_y = (iVal o Bits28) y
   in
   bus32_num_append
     (@bit28_val:word28.
        (iVal o Bits28) bit28_val = ival_x modulo_28_Add ival_y))
```

The **SUB28** and **ADD28** functions are the specifications for the operations performed by the low level component, and is the same as is specified at the top level. The use of the same "modulo_28_Add" operation in defining the corresponding operation at both levels makes this explicit. The reason the operations differ is that they operate on different data types; the top level has integer operands while the register transfer level has 28-bit operands (of type *:word28*). The proof of correctness of the arithmetic components of the chip is thus part of the problem of relating the RT and lower levels of description. The relation between the data types and abstraction functions and addition may be charted as follows.

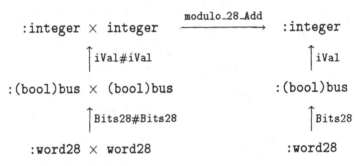

Addition on *:word28* arguments has been defined as returning that number record whose interpretation as an integer is the same as the integer representing the sum of the interpretations of the operands. Thus the ADD28 function could be drawn as an arrow across the bottom of the diagram from the types :word28 × word28 to, not the type :word28 but rather the type :word32 obtained by applying the function bus32_num_append to it.

Despite both M_Add and ADD28 being defined in terms of the same underlying modulo_28_Add operation, when we try to relate the two operations, we find it necessary to prove several properties about the Val and iVal abstractions, leading to a proof of the unique existence of :word28 values for each integer returned by the norm28 function.

The output of the ALU is gated onto the bus under the control of the ralu signal. When no ALU operation is selected, the value computed is not significant. The definition uses implication to define the computed output only when one of the alu control signals is asserted. If none of the control signals is asserted, the value of alu is undefined. Likewise, if more than one control line is asserted, the computed output is also unimportant. This is expressed in the specification by the use of a one_asserted predicate on the control lines which implies the desired behaviour. This predicate states that if one of the signals is asserted, the remainder are not.

```
ALU replcar replcdr sub add dec mul div rem
    setbit30 setbit31 resetbit30 resetbit31
    ralu arg y2 bus alu =
(one_asserted_12 replcar replcdr sub add dec mul div rem
    setbit30 setbit31 resetbit30 resetbit31 ==>
(!tₘ.
 let bus28 = (atom_bits (bus tₘ))
 in
 let arg28 = (atom_bits (arg tₘ))
 in
 ((replcar tₘ    ==> (alu tₘ = REPLCAR(arg tₘ)(y2 tₘ))) ∧
  (replcdr tₘ    ==> (alu tₘ = REPLCDR(arg tₘ)(y2 tₘ))) ∧
  (sub tₘ        ==> (alu tₘ = SUB28     arg28     bus28)) ∧
  (add tₘ        ==> (alu tₘ = ADD28     arg28     bus28)) ∧
  (dec tₘ        ==> (alu tₘ = DEC28     arg28))            ∧
  (mul tₘ        ==> (alu tₘ = DEC28     arg28))            ∧
  (div tₘ        ==> (alu tₘ = DEC28     arg28))            ∧
  (rem tₘ        ==> (alu tₘ = DEC28     arg28))            ∧
  (setbit31 tₘ   ==> (alu tₘ = SETBIT31(arg tₘ)))          ∧
  (setbit30 tₘ   ==> (alu tₘ = SETBIT30(arg tₘ)))          ∧
  (resetbit31 tₘ ==> (alu tₘ = RESETBIT31(arg tₘ)))        ∧
  (resetbit30 tₘ ==> (alu tₘ = RESETBIT30(arg tₘ)))))))
∧
(!tₘ. ralu tₘ ==> (bus tₘ = alu tₘ)))
```

The datapath DP, defined in Figure 4.9, consists of the conjunction of subcomponents, closely resembling the diagram in Figure 3.2. The bus is expressed as the wiring together of the components. The CAR register is unique in having the input connected to the *car* field of the bus, and the output to the *cdr* field. All other 14 bit registers have both input and output connected to the *cdr* field of the bus.

4.4.3 The control unit specification

The CU differs most from the lower level. The layout hierarchy is replaced by one that better reflects its functionality, using three components for its definition: a *state register*, a microcode *ROM*, and a *decode* section. Conjoining the definitions of these three parts gives a typical finite state machine behaviour, defining the next state and current output values as a function of the current state and current input values.

```
DP bus_bits mem_bits  SYS_Clocked  rmem  mar wmar rmar
   rnum rnil rtrue rfalse  s ws rs  e we re  c wc rc   d wd rd
   free wfree rfree  parent wparent rparent  root wroot rroot
   y1 wy1 ry1  x1 wx1 rx1  x2 wx2 rx2  y2 wy2 ry2  rcons
   car wcar rcar  atomflag bit30flag bit31flag zeroflag
   nilflag eqflag leqflag  arg warg rarg  buf1 wbuf1 rbuf1
   buf2 wbuf2 rbuf2  replcar replcdr sub add dec mul div rem
   setbit30 setbit31 resetbit30 resetbit31  ralu =
?alu:num->word32.
   let a_bus = (λt_m. car_bits (bus_bits t))
   and d_bus = (λt_m. cdr_bits (bus_bits t))
   in
   ((READ_MEM mem_bits rmem bus_bits)                         ∧
    (MAR a_bus d_bus SYS_Clocked wmar rmar mar)               ∧
    (NUM    rnum   a_bus d_bus)                               ∧
    (Nil    rnil   d_bus)                                     ∧
    (TRUE   rtrue  d_bus)                                     ∧
    (FALSE rfalse d_bus)                                      ∧
    (S_reg       d_bus SYS_Clocked ws        rs      s)       ∧
    (E_reg       d_bus SYS_Clocked we        re      e)       ∧
    (C_reg       d_bus SYS_Clocked wc        rc      c)       ∧
    (D_reg       d_bus SYS_Clocked wd        rd      d)       ∧
    (FREE_reg    d_bus SYS_Clocked wfree     rfree   free)    ∧
    (PARENT_reg d_bus SYS_Clocked wparent rparent parent) ∧
    (ROOT_reg    d_bus SYS_Clocked wroot     rroot   root)    ∧
    (Y1_reg      d_bus SYS_Clocked wy1       ry1     y1)      ∧
    (X1_reg      d_bus SYS_Clocked wx1       rx1     x1)      ∧
    (X2_reg      d_bus SYS_Clocked wx2       rx2     x2)      ∧
    (Cons x1 x2 rcons bus_bits)                               ∧
    (CAR_reg a_bus d_bus SYS_Clocked wcar rcar car)           ∧
    (FLAGSUNIT bus_bits arg atomflag bit30flag bit31flag
               zeroflag nilflag eqflag leqflag)               ∧
    (Y2_reg      d_bus SYS_Clocked wy2       ry2     y2)      ∧
    (ARG_reg bus_bits SYS_Clocked warg     rarg    arg)      ∧
    (ALU replcar replcdr sub add dec mul div rem
         setbit30 setbit31 resetbit30 resetbit31
         ralu arg y2 bus_bits alu)                            ∧
    (BUF1_reg alu bus_bits SYS_Clocked wbuf1 rbuf1 buf1)  ∧
    (BUF2_reg alu bus_bits SYS_Clocked wbuf2 rbuf2 buf2))
```

Figure 4.9: Definition of the DP Component

The control unit utilizes several fixed length word types, :**word27** for the microinstruction word (*ROM* output), :**word4** for *test* and *alu* fields of a microinstruction, :**word5** for *read* and *write* fields, and :**word9** for the *A* field and the micro pc (*ROM* input). Field extractor functions include: **Read_field**, **Write_field**, **Alu_field**, **Test_field**, and **A_field**. An increment function for :**word9** values is used to find the next address value for sequential microcode execution. It is defined in two parts, the **inc** function operates on boolean buses, and presents a very implementation-like view. It returns both an incremented *bus* as well as a propagate flag. **Inc9** works on values of type :**word9**, first extracting the *bus* value, then applying the **inc** function, taking only the bus part of the returned value, and casting it as a :**word9** result.

```
(inc ((Wire b):(bool)bus) = (Wire (~b),b))  ∧
(inc (Bus b bus) =
   let (bs,fg) = (inc bus)
   in
   (Bus (fg => ¬b | b) bs, (fg ∧ b)))

Inc9:word9->word9 = Word9 o FST o inc o Bits9
```

The state register component for the control unit includes not only the *mpc* register, but a 4-deep *stack* for microcode subroutine calls as well. The definition of the **S_latch9** differs slightly from the **MPC9** register, lacking any provision for resetting. They are implemented at the low level by altering the clocking signal for the first (master) latch of the pair comprising the register. This signal is the AND of ϕ_A and a load signal, which is the OR of the *pop* and *push* control signals. This restricts latching of new values to only those times when a stack command is issued. At other times, the value in the master latch stays unchanged, and the slave latch reloads this same value during ϕ_B.

```
S_latch9 SYS_clk (load:num->bool)(in:num->word9)(out:num->word9) =
!t_m. out (t_m+1) =
        SYS_clk t_m => (load t_m => in t_m | out t_m)
                     | out t_m

STATE_REG SYS_clk load
          (next_mpc,next_s0,next_s1,next_s2,next_s3)
          (    mpc,     s0,     s1,     s2,     s3) =
(MPC9    SYS_clk          next_mpc mpc) ∧
(S_latch9 SYS_clk load next_s0 s0) ∧
(S_latch9 SYS_clk load next_s1 s1) ∧
(S_latch9 SYS_clk load next_s2 s2) ∧
(S_latch9 SYS_clk load next_s3 s3)
```

The microcode ROM in the SECD chip is a $9 \times 400 \times 27$ ROM, one address of which is unused for the microcode and returns all zeros. The desired HOL representation of this device is a function of type :word14→word27, which returns the :word27 equivalent of the 27 bit microinstruction for each of the 399 microcode :word9 address inputs.

Limiting the function specification to only those microcode addresses used is preferable to total specification, since the latter would limit acceptable implementations to those returning some arbitrarily selected value for unused address inputs. By proving the existence of a value satisfying a property "P", a new constant can be created that has this property. The desired property is given as the conjunction of equations giving a function value at 399 distinct :word9 constants.

```
(f #000000000 = #000010110000100000000000000) ∧
(f #000000001 = #000010111000100000000000000) ∧ ....
```

A function satisfying this property can be built as a binary decision tree, with :word27 records at the leaves, and arbitrary values on branches selected by unused addresses. The path through the tree is determined by the 9 bits of the :word9 input. The function is designed to terminate when reaching a leaf, so the tree is minimal, thereby reducing the term size in subsequent proofs. This function is evaluated at each of 399 :word9 constants, and the conjunction of the theorems so generated proves it is a witness to the existence of a function with the desired property. The ML function *new_specification* creates a new constant, and returns a theorem of the form ⊢ P *new_constant_name*.

```
ROM_fun_thm =
⊢ (ROM_fun #000000000 = #000010110000100000000000000) ∧
  (ROM_fun #000000001 = #000101111000100000000000000) ∧
  (ROM_fun #000000010 = #001010110000100000000000000) ∧
  ...
  (ROM_fun #110001100 = #000000000000000000001000101) ∧
  (ROM_fun #110001101 = #101100110000110100000100001) ∧
  (ROM_fun #110001110 = #000000000110000000000000000)
```

The only difficulty in this approach is proving the large number of theorems evaluating the binary decision tree, within the constraint of system imposed limitations, particularly swap space. The swap space limit was boosted to 30 megabyte before the original proof could be completed (process size reached approximately 23 Mbyte). An intermediate form of the microcode image used to generate the microcode ROM layout is used to generate the set of 27-bit values in the tree, as well as the 9-bit addresses.

The CU_DECODE component has 61 outputs: 12 ALU control signals, 18 write signals, one of which controls the bidirectional i/o pads, 23 read signals, 2 state flags (flag0,flag1), a stack control signal (push_or_pop), the input for the MPC9 state register (nextmpc), and inputs for each of the 4 stack registers (next_s0...next_s3); 15 inputs: button, the current mpc value, the value in each stack cell (s0...s3), the 27 bit ROM output (rom_out), the machine instruction input (opcode), and 7 status flags from the datapath. There are no hidden lines as such, but the definition makes extensive use of let bindings to simplify the term. The let bound terms represent the incremented mpc value, the 5 fields of the ROM output, flags for each of the 3 states identified by the state flags, and complex logical values that control the selection of *next* state outputs for the MPC9 and stack. The entire control unit definition is given in Figure 4.11.

```
CU_DECODE  button mpc  s0 s1 s2 s3  rom_out  opcode
  atomflag bit30flag bit31flag zeroflag nilflag eqflag leqflag
  flag0 flag1  nextmpc next_s0 next_s1 next_s2 next_s3
  push_or_pop  ralu rmem rarg rbuf1 rbuf2 rcar rs re rc rd rmar
  rx1 rx2 rfree rparent rroot ry1 ry2 rnum rnil rtrue rfalse rcons
  write_bit wmem_bar warg wbuf1 wbuf2 wcar ws we wc wd wmar wx1 wx2
  wftee wparent wroot wy1 wy2 dec add sub mul div rem
  setbit30 setbit31 resetbit31 replcar replcdr resetbit30 =
  !t_m.
  let mpc_plus_1  = Inc9 (mpc t_m) in
  let write_bits  = Write_field (rom_out t_m) in
  let read_bits   = Read_field (rom_out t_m) in
  let alu_bits    = Alu_field  (rom_out t_m) in
  let test        = Test_field (rom_out t_m) in
  let A_address   = A_field    (rom_out t_m) in
  let idle_state  = (mpc t_m = #000010110)  in
  let error_state = (mpc t_m = #000011000)  in
  let toc_state   = (mpc t_m = #000101011)  in
  let selA        =
   ( (test = #0001)                          V ((test = #0011) ∧ bit30flag t_m)V
    ((test = #0100) ∧ bit31flag t_m) V ((test = #0101) ∧ eqflag t_m)    V
    ((test = #0110) ∧ leqflag  t_m) V ((test = #0111) ∧ nilflag t_m)    V
    ((test = #1000) ∧ atomflag t_m) V ((test = #1001) ∧ zeroflag t_m)   V
    ((test = #1010) ∧ button   t_m) V (test = #1011))       in
  let pop       = (test = #1100)     in
  let push      = (test = #1011)     in
  let sel0p     = (test = #0010)
  in
  ((flag0 t_m     = idle_state V error_state)                     ∧
   (flag1 t_m     = toc_state  V error_state)                     ∧
   (nextmpc t_m   = ( (selA)   =>  A_address
                    | (pop)    =>  s0 t_m
                    | (sel0p)  =>  opcode t_m | mpc_plus_1)))      ∧
   ((next_s0 t_m, next_s1 t_m, next_s2 t_m, next_s3 t_m) =
       push => (mpc_plus_1, s0 t_m, s1 t_m, s2 t_m)
     | pop  => (s1 t_m, s2 t_m, s3 t_m, #000000000)
              | (s0 t_m, s1 t_m, s2 t_m, s3 t_m))                 ∧
    (push_or_pop t_m = push V pop)                                ∧
    (ralu     t_m = (read_bits = #00001))                         ∧
    (rmem     t_m = (read_bits = #00010))                         ∧
    (rarg     t_m = (read_bits = #00011))                         ∧
    (rbuf1    t_m = (read_bits = #00100))                         ∧
    (rbuf2    t_m = (read_bits = #00101))                         ∧
    (rcar     t_m = (read_bits = #00110))                         ∧
    (rs       t_m = (read_bits = #00111))                         ∧
    (re       t_m = (read_bits = #01000))                         ∧
    (rc       t_m = (read_bits = #01001))                         ∧
    (rd       t_m = (read_bits = #01010))                         ∧
    (rmar     t_m = (read_bits = #01011))                         ∧
```

```
(rx1         t_m = (read_bits = #01100))              ∧
(rx2         t_m = (read_bits = #01101))              ∧
(rfree       t_m = (read_bits = #01110))              ∧
(rparent     t_m = (read_bits = #01111))              ∧
(rroot       t_m = (read_bits = #10000))              ∧
(ry1         t_m = (read_bits = #10001))              ∧
(ry2         t_m = (read_bits = #10010))              ∧
(rnum        t_m = (read_bits = #10011))              ∧
(rnil        t_m = (read_bits = #10100))              ∧
(rtrue       t_m = (read_bits = #10101))              ∧
(rfalse      t_m = (read_bits = #10110))              ∧
(rcons       t_m = (read_bits = #10111))              ∧

(write_bit   t_m = ¬(write_bits = #00001))            ∧
(wmem_bar    t_m = ¬(write_bits = #00001))            ∧
(warg        t_m = (write_bits = #00010))             ∧
(wbuf1       t_m = (write_bits = #00011))             ∧
(wbuf2       t_m = (write_bits = #00100))             ∧
(wcar        t_m = (write_bits = #00101))             ∧
(ws          t_m = (write_bits = #00110))             ∧
(we          t_m = (write_bits = #00111))             ∧
(wc          t_m = (write_bits = #01000))             ∧
(wd          t_m = (write_bits = #01001))             ∧
(wmar        t_m = (write_bits = #01010))             ∧
(wx1         t_m = (write_bits = #01011))             ∧
(wx2         t_m = (write_bits = #01100))             ∧
(wfree       t_m = (write_bits = #01101))             ∧
(wparent     t_m = (write_bits = #01110))             ∧
(wroot       t_m = (write_bits = #01111))             ∧
(wy1         t_m = (write_bits = #10000))             ∧
(wy2         t_m = (write_bits = #10001))             ∧

(dec         t_m = (alu_bits = #0001))                ∧
(add         t_m = (alu_bits = #0010))                ∧
(sub         t_m = (alu_bits = #0011))                ∧
(mul         t_m = (alu_bits = #0100))                ∧
(div         t_m = (alu_bits = #0101))                ∧
(rem         t_m = (alu_bits = #0110))                ∧
(setbit30    t_m = (alu_bits = #0111))                ∧
(setbit31    t_m = (alu_bits = #1000))                ∧
(resetbit31  t_m = (alu_bits = #1001))                ∧
(replcar     t_m = (alu_bits = #1010))                ∧
(replcdr     t_m = (alu_bits = #1011))                ∧
(resetbit30  t_m = (alu_bits = #1100)))
```

Figure 4.10: The CU_DECODE definition

```
CU SYS_Clocked button mpc s0 s1 s2 s3 opcode atomflag
  bit30flag bit31flag zeroflag nilflag eqflag leqflag
  flag0 flag1 ralu rmem rarg rbuf1 rbuf2 rcar rs re rc rd
  rmar rx1 rx2 rfree rparent rroot ry1 ry2 rnum rnil rtrue
  rfalse rcons write_bit bidir warg wbuf1 wbuf2 wcar
  ws we wc wd wmar wx1 wx2 wfree wparent wroot wy1 wy2
  dec add sub mul div rem setbit30 setbit31 resetbit31
  replcar replcdr resetbit30
  =
? (rom_out:num->word27) (nextmpc:num->word9) (next_s0:num->word9)
  (next_s1:num->word9)  (next_s2:num->word9) (next_s3:num->word9)
  (push_or_pop:num->bool).
(STATE_REG SYS_Clocked push_or_pop
  (nextmpc, next_s0, next_s1, next_s2, next_s3)
  (   mpc,      s0,     s1,     s2,     s3))
∧
(ROM_t mpc rom_out)
∧
(CU_DECODE button  mpc  s0 s1 s2 s3  rom_out  opcode
  atomflag bit30flag bit31flag zeroflag nilflag eqflag leqflag
  flag0 flag1  nextmpc next_s0 next_s1 next_s2 next_s3
  push_or_pop  ralu rmem rarg rbuf1 rbuf2 rcar rs re rc rd rmar
  rx1 rx2 rfree rparent rroot ry1 ry2 rnum rnil rtrue rfalse rcons
  write_bit bidir warg wbuf1 wbuf2 wcar ws we wc wd wmar wx1 wx2
  wfree wparent wroot wy1 wy2 dec add sub mul div rem
  setbit30 setbit31 resetbit31 replcar replcdr resetbit30)
```

Figure 4.11: Definition of the CU Component

4.4.4 The padframe

The padframe includes input, output, and bidirectional pads. The input and output pads specify equality at all times of the pairs of nodes connected by the pads. The bidirectional pads drive the value at bus_bits onto bus_pins when writing off-chip, and when reading from off-chip (also the default) drive the value at bus_pins onto mem_bits, which is ported back to the bus through the READ_MEM unit described earlier. The subcomponents are not defined separately, but the functionality is specified in the PF definition in Figure 4.12.

```
PF button flag0 flag1 bidir write_bit rmem          % chip side %
      bus_bits mem_bits mar_bits
      button_pin flag0_pin flag1_pin                % pin side  %
      write_bit_pin rmem_pin bus_pins mar_pins =
    !t_m.
    (button t_m        = button_pin t_m)       ∧
    (flag0_pin t_m     = flag0 t_m)            ∧
    (flag1_pin t_m     = flag1 t_m)            ∧
    (write_bit_pin t_m = write_bit t_m)        ∧
    (rmem_pin t_m      = rmem t_m)             ∧
    (bidir t_m => (mem_bits t_m = bus_pins t_m)     % read mem  %
           |  (bus_pins t_m = bus_bits t_m)) ∧    % write mem %
    (mar_pins t_m      = mar_bits t_m)
```

Figure 4.12: Definition of the PF Component

4.4.5 Composing the whole

The SECD chip is the composition of the CU, DP, and PF. The parameters to the definition include the SYS_Clocked signal; control unit state values mpc, s0, s1, s2 and s3; datapath state values s, e, c, d, free, x1, x2, y1, y2, car, root, parent, buf1, buf2 and arg; input button_pin; state outputs flag0_pin and flag1_pin; memory interface outputs write_bit_pin, rmem_pin and mar_pins; and the bidirectional memory bus bus_pins. A multitude of signals are hidden, including the *read, write* and *alu* control signals generated by the CU, status flag signals, and on-chip signals connected through the pads, including button, flag0, flag1, bus_bits, mem_bits and mar_bits. Notice that the CU component has the argument (Opcode arg) as the machine instruction opcode. The Opcode function selects the lower 9 bit word from the arg value.

The external memory (RAM) is represented only at the top level. It has only two operations given by the Fetch and Store functions, and its current state is held in the first parameter memory. The conjunction of the chip SECD and an SRAM memory constitutes the definition of the complete system shown in Figure 4.14.

```
SECD SYS_Clocked mpc s0 s1 s2 s3
   button_pin flag0_pin flag1_pin
   write_bit_pin rmem_pin bus_pins mar_pins
   s e c d free x1 x2 y1 y2 car root parent
   buf1 buf2 arg
   =
 ? button atomflag bit30flag bit31flag zeroflag nilflag eqflag
   leqflag  flag0 flag1   ralu rmem rarg rbuf1 rbuf2 rcar
   rs re rc rd rmar rx1 rx2 rfree rparent rroot ry1 ry2
   rnum rnil rtrue rfalse rcons
   write_bit bidir warg wbuf1 wbuf2 wcar ws we wc wd wmar wx1 wx2
   wfree wparent wroot wy1 wy2 dec add sub mul div rem
   setbit30 setbit31 resetbit31 replcar replcdr resetbit30
   bus_bits mem_bits mar_bits .

   (CU SYS_Clocked button mpc s0 s1 s2 s3 (Opcode arg) atomflag
      bit30flag bit31flag zeroflag nilflag eqflag leqflag
      flag0 flag1 ralu rmem rarg rbuf1 rbuf2 rcar rs re rc rd
      rmar rx1 rx2 rfree rparent rroot ry1 ry2 rnum rnil rtrue
      rfalse rcons write_bit bidir warg wbuf1 wbuf2 wcar
      ws we wc wd wmar wx1 wx2 wfree wparent wroot wy1 wy2
      dec add sub mul div rem setbit30 setbit31 resetbit31
      replcar replcdr resetbit30)
   ∧
   (DP bus_bits mem_bits  SYS_Clocked  rmem  mar_bits wmar rmar
      rnum rnil rtrue rfalse  s ws rs  e we re  c wc rc  d wd rd
      free wfree rfree  parent wparent rparent  root wroot rroot
      y1 wy1 ry1  x1 wx1 rx1  x2 wx2 rx2  y2 wy2 ry2  rcons
      car wcar rcar  atomflag bit30flag bit31flag zeroflag
      nilflag eqflag leqflag  arg warg rarg  buf1 wbuf1 rbuf1
      buf2 wbuf2 rbuf2  replcar replcdr sub add dec mul div rem
      setbit30 setbit31 resetbit30 resetbit31  ralu)
   ∧
   (rt_PF button flag0 flag1 bidir write_bit rmem
      bus_bits mem_bits mar_bits
      button_pin flag0_pin flag1_pin
      write_bit_pin rmem_pin bus_pins mar_pins)
```

Figure 4.13: RT level SECD definition

```
Fetch4 (addr:word14)(bus:word32)(memory:word14->word32) =
  (bus = memory addr)

Store14 (addr:word14)(bus:word32)(memory:word14->word32) =
  (λa. (a = addr) => bus | memory a)

SRAM (memory:num->(word14->word32))(W_bar:num->bool)
     (G_bar:num->bool)(addr:num->word14)(bus:num->word32) =
!tₘ. (memory(SUC tₘ) =
        ((¬W_bar tₘ)
         => Store14(addr tₘ)(bus tₘ)(memory tₘ)
          | memory tₘ)) ∧
      (W_bar tₘ ∧ G_bar tₘ ==>
              Fetch14(addr tₘ)(bus tₘ)(memory tₘ)))
```

```
SYS memory SYS_Clocked mpc s0 s1 s2 s3
    button_pin flag0_pin flag1_pin
    write_bit_pin rmem_pin bus_pins mar_pins
    s e c d free x1 x2 y1 y2 car root parent
    buf1 buf2 arg
  =
(SECD SYS_Clocked mpc s0 s1 s2 s3
    button_pin flag0_pin flag1_pin
    write_bit_pin rmem_pin bus_pins mar_pins
    s e c d free x1 x2 y1 y2 car root parent
    buf1 buf2 arg)
∧
(SRAM memory write_bit_pin rmem_pin mar_pins bus_pins)
```

Figure 4.14: Definition of the SECD System

4.5 Relating the Levels

In order to relate the behaviours of the RT and top levels, we must establish a correspondence between the different data types and the different time granularities. This is done by defining abstractions

between the object at the RT level and the corresponding object at
the top level.

4.5.1 Memory abstraction

Abstracting from the SRAM memory of the RT level to the abstract
memory type maintains the mark and field bits unchanged, and maps
the 28 bit field to the appropriate *cons*, *integer*, or *symbol* record
based on the record type bits. The memory abstraction function is
defined in two steps: first a function maps records in the codomain
of the SRAM memory function into the codomain of the representative
type for abstract memories, and then ABS_mfsexp_mem is applied to
this function composed with the SRAM memory:

```
Mem_Range_Abs:word32 -> ((bool#bool)#((word14#word14)+atom)) w =
  ((mark_bit w),(field_bit w)),
  ( (is_symbol w) => INR(Symb(Val(Bits28(atom_bits w))))
  | (is_number w) => INR(Int(iVal(Bits28(atom_bits w))))
                     | INL(car_bits w, cdr_bits w))

mem_abs(M:word14->word32) = ABS_mfsexp_mem(Mem_Range_Abs o M)
```

The Mem_Range_Abs function is total, and the unused record type
(#01) gets mapped to the *cons* type record of the abstract memory.
This ensures that the result of composing it with an implementation
memory always returns an object in the representative type of the
abstract memory. This is vital, since it is often the case that we will
apply REP_mfsexp_mem to (mem_abs M) at some value v, and when
we unfold the definition of mem_abs, we have the term

REP_mfsexp_mem (ABS_mfsexp_mem(Mem_Range_Abs o M)) v.

The REP_ composed with the ABS_ function is an identity only on
objects in the representative type of :(word14,atom)mfsexp_mem.
Thus we can simplify the expression to Mem_Range_Abs (M v). It
further ensures the desirable property that every non-atomic record
is a *cons* type record. The machine's correct behaviour should ensure
that the unused record type never occurs, but the initial memory
configuration could admit unused cells which were not placed in the

free list, and about which we have no knowledge of the value of the
record type bits.

4.5.2 Major states abstraction

The state of the system at the RT level comprises the contents of all
the data registers, the memory, and the control unit state register. At
the top level, the state comprises selected register contents, the abstract memory, and a major state value. The major state values are
idle, top_of_cycle, error0 and error1; they are defined as representing pairs of bools. An abstraction is defined from the value in the
MPC9 register to these four states. The function is totally defined, but
returns an arbitrary value from any microinstruction address which
does not correspond to being in a major state. Alternatively, and
perhaps preferably, a partially defined but total function could be
used.

```
state_abs mpc =
   (mpc = #000010110) => idle
 | (mpc = #000011000) => error0
 | (mpc = #000011010) => error1
 | (mpc = #000101011) => top_of_cycle
 |                        (@stat.F)
```

A predicate is also defined to hold when the mpc value is one which
corresponds to a major state. This is used for the temporal abstraction which follows.

```
is_major_state mpc t_m =
   (mpc t_m = #000010110) ∨
   (mpc t_m = #000011000) ∨
   (mpc t_m = #000101011) ∨
   (mpc t_m = #000011010)
```

4.5.3 Temporal abstraction

The coarsest grain of time used to describe the system corresponds to
the points when the system is in a major state. We map between this
coarse granularity and the medium grain at points of time when the

predicate **is_major_state** mpc holds. The mapping is not a linear function, since the number of cycles needed to execute any machine instruction varies, and can vary between executions of the same instruction. The latter differences arise due to garbage collection calls during instruction execution, as well as varying search distances required to load values from the environment. The definitions of the mapping functions are based on Melham's work in [43].

```
Next (t1:num) (t2:num) (f:num->bool) =
(t1 < t2) ∧ (f t2) ∧ !t.(t1 < t) ∧ (t < t2) ==> ¬f t

(IsTimeOf 0 (f:num->bool) (t:num) =
        f t ∧ !t'.(t'<t) ==> ¬f t') ∧
(IsTimeOf (SUC n) f t = ?t'.IsTimeOf n f t' ∧ Next t' t f)

TimeOf (f:num->bool) (n:num) = @t.IsTimeOf n f t

(s:num->*) when (p:num->bool) = λn:num. s (TimeOf p n)

Inf f =  !t. ?t'.  (t<t') ∧ (f t')
```

Figure 4.15: Temporal Abstraction Function Definitions

Next defines a predicate which holds when t2 is the next time that f holds after t1. **TimeOf** when applied to a predicate (such as **is_major_state**) and a point of coarse grain time n_c, will return the corresponding time at the medium granularity, say t_m. In the case where the mapping is finite, there would exist a maximum value above which the function would be undefined. Such a function would be partial, and cannot be directly defined within the logic of HOL, and thus its definition is based on a *relation* **IsTimeOf**. The infix operator **when** can then abstract a signal sampled at the medium granularity to one sampled at the coarse granularity of time.

The definition of the abstraction functions may only be used when the value at the medium granularity is well defined, and for this purpose the predicate must be shown to be true infinitely often, as captured by the predicate: **Inf is_major_state**.

4.6 Summary

This chapter has presented two complete levels of formal specification of the SECD system, the top level specification and the register transfer level implementation. Important issues at the lowest level that impact the higher levels were discussed. Lastly, three important abstractions used to relate the different levels were presented.

The top level specification needs to express transformations to S-expression data objects. Their representation is complicated by the destructive operation used to create circular environment components, and the fact that the expressions can be shared by the different registers, so that a destructive operation to the E stack can also affect the S stack. An abstract memory data type is defined in which the data structures can be embedded. A set of operations on abstract memory data objects express transformations to the data structures. This technique tries to provide a clearer specification than would have been possible by using a lower level view of a memory and system state. Additional motivation is the complexity of the machine instruction transitions involving multiple updates to memory, giving rise to considerably more complexity than previous microprocessor specification and verification examples, which typically are limited to single memory updates in any transition. It is also hoped that it provides a more suitable specification to interface with software semantic specifications.

The discussion of the low level specification focussed on fundamental issues of clock operation and initialisation, and the definition of primitive memory devices. The notion of fine grain time integrates the operation of two distinct clocks, where units of time relate to the advance of either clock.

It is our belief that the register transfer level is an appropriate lower level of representation for verification of VLSI designs. Most of the primitive components used should become available as verified library components as the field of verification matures. The control part of the design is expressed more succinctly than the lower level, but relating the two representations is relatively simple. This representation closely resembles the informal Concrete RTL model created in designing the chip. The closeness suggests that formal specification may integrate into the design process quite naturally, providing a useful tool for expressing the information that the designers actually use.

The absence of a definition for the garbage collection function illustrates how a formal specification may be incrementally developed in a top down manner, when not all parts have been fully elaborated. The precise nature of the garbage collection function is not defined, but its *type* can be established, knowing simply that it makes some changes to a memory. By defining a constant of the appropriate type, filling in the detail can be deferred.

One of the more important and difficult parts of the specification concerns the temporal representations, and abstractions relating the different time granularities. Defining precisely how the clocks should operate, and determining how to represent the behaviour at the next higher level are fundamental issues in writing the operating specification for the SECD chip [26], which formed the basis for the design of a chip controller board and associated software [54].

A great deal of computational effort results from the use of a well-defined $word_n$ type. Previous HOL proofs by Cohn [12], Joyce[34] and others relied on insecurely introduced constants, or infinite vectors of which values beyond a certain size were assigned an "arbitrary" value [35]. The advantages of the well defined types used herein include the higher confidence level from avoiding the arbitrary introduction of new axioms, and the more natural definition of subfield functions on $word_n$ values. This latter advantage is not insignificant, since the resulting specifications are much more clear, and easier to get right the first time.

The sheer size of this system makes writing the specification a daunting task. In retrospect, some of the most challenging aspects of the project involved the development of clear and concise definitions in the specification. Definitions were repeatedly revised, as the design itself and issues in verifying the design became better understood. The simplicity of many definitions belie the amount of care taken in their design.

The organization of the HOL theory hierarchy in which the system is defined bears mentioning. Whenever a definition is altered, all theories that inherit the definition need rebuilding. A change to a data type theory high in the hierarchy could require many hours of updating of HOL theories. The maintenance was aided considerably by keeping a well documented makefile, and overnight rebuilds. Careful design of the theory hierarchy and separation of dependencies can save a lot of time. It should be noted that even at the last stages of the proof, changes high in the theory hierarchy were made, as clumsy definitions done at the early stages of the project were replaced.

To give some scale of the task of defining the complete system at two levels, we report the number of primitive inferences recorded by the HOL system. The introduction of new data types and function definitions is often accompanied by at least a partial axiomatization. Data type definitions and associated theorems take over 50,000 primitive inferences. These high numbers result from recursive definitions on the $word_n$ data types. Specifications require over 31,000 primitive inferences, a substantial portion of which is expended in the specification of the ROM function. Definition of abstraction functions and some associated proofs add up to over 35,000 primitive inferences.

Exploring further

By providing the HOL sources for the SECD specification, we hope to provide a useful source of teaching examples which involve dealing with a nontrivial design. Some suggested areas to explore include:

1. Complete a full low level definition of the chip subcomponents.

2. Alter the RTL specification to include the shift register components, noticing that the signals trapped include some which are presently internal to the control unit.

3. Specify constraints on the shift register control inputs, integrating these with a full set of clocking constraints.

4. Specify a top level behaviour for test mode of the system operation.

5. Specify the behaviour of the garbage collector. Consider other data type representations for the top level specification that might make this problem easier.

Chapter 5

Verification of the SECD Design

A proof of correctness relates two levels of description of a design, proving that, subject to stated constraints, the behaviour of the lower level of description ensures the behaviour specified at the higher level. Parameters to the higher level description are abstractions of the lower level ones, with regard to temporal granularity and data type. The lower level can be considered an *implementation* of the higher level *specification*. Using this terminology, the goal of a proof has the form:

constraints ⊃
    implementation (state) (inputs) (outputs) ⊃
      specification (abs o state) (abs o inputs) (abs o outputs)

It is desirable that a series of such correctness proofs relate the lowest level description (that closest to the physical device), to the highest level (that which comes closest to the designer's intention).

This chapter will describe the proof of correctness relating the top and RT levels, under normal mode of operation, and exclusive of garbage collection. The abstractions on signal and state values must map between two granularities of time, as well as different data types, particularly for the memories. The register transfer definition includes a memory function with simple fetch and store operations only, while the top level uses the abstract memory data type. The task of the verification is to show that the sequence of operations performed at the register transfer level commutes with the specification transition at the more abstract top level.

$$\text{System State} \xrightarrow[\text{transition}]{\text{specification}} \text{System State'}$$

$$\text{abstraction} \uparrow \qquad\qquad \uparrow \text{abstraction}$$

$$\text{RTL State} \xrightarrow[\substack{\text{RTL} \\ \text{transitions}}]{*} \text{RTL State'}$$

This level of the proof is largely concerned with control operation, almost completely ignoring arithmetic and logical comparison operation semantics. A proof of correctness relating the lowest level and register transfer level must address these issues, defining precisely the implementation of the arithmetic operations. We choose to focus upon the verification of the control operation because this is central to the nature of a microprocessor that distinguishes it from less complex devices. There is abundant literature in this field already using arithmetic logic unit implementations as verification subjects.

The constraints on the scope of the proof are addressed first, then the form of the correctness goal and the approach used in achieving the proof are given. Following that, each major stage in the proof is outlined, using the LDF transition as the running example.

5.1   Constraints

Constraints limiting the scope of the proof to normal mode of operation and excluding garbage collection need to be formalised. Additionally, the proof is limited to valid programs, which expresses the pattern matching in the left side of transitions defining the informal machine (Table 2.3), and to properly configured memories with respect to reserved words, free list configurations and the value of the *mark* and *field* bits. Lastly, two assumptions about the decrement operation are listed for the reader's benefit. The assumptions are discharged late in the proof, but they will appear in many intermediate theorems. The constraints are as follows.

clock_constraint The chip is running in normal operating mode (*i.e.* the shift registers are not operating).

---

clock_constraint SYS_Clocked = !t_m. SYS_Clocked t_m

---

valid_program_constraint There are two parts to the constraint shown in Figure 5.1. The first applies when a program has been installed in memory and the machine is to begin execution. The reserved memory location at **NUM_addr** must point to a data structure with a program part, a data part, and a free list. The values of these objects are not themselves constrained at this point.

The second part applies at the start of execution of any machine instruction. The state of the machine must pattern match with the left side of the abstract machine transition for one of the 18 implemented machine instructions, or have a problem loaded in memory when the machine is directed to start computation. The constraint gives a quite detailed specification of the type and in some cases the value of each record in memory involved in the pattern match. The constraint on record types is necessary for the abstraction from simple to abstract memory types. This **valid_program_constraint** must be assured by the correctness of the compilation process.

The component constraining the arguments for the **LD** instruction requires further explanation. First it is required that the argument is a cons cell, and each branch points to a number cell. The integer values represented by these cells are non-negative, and furthermore, the environment has a value in the position indicated; *i.e.* it is composed of at least m lists, and the mth list has at least n elements.

reserved_words_constraint There are three reserved locations in memory that contain the symbolic constants for **NIL**, **T**, and **F**. In view of the role of an outside agency in downloading a problem into memory, likely while the chip is in the *idle* state, the constraint only requires the reserved words to be in place once it is ready to begin executing a program.

```
valid_program_constraint memory mpc button_pin s e c d =
!t_m.
(((state_abs(mpc t_m) = idle) ∧ button_pin t_m) ==>
   (is_cons(memory t_m NUM_addr)) ∧
   (is_cons(memory t_m(car_bits(memory t_m NUM_addr)))))
∧
((state_abs(mpc t_m) = top_of_cycle) ==>
   let head_c = memory t_m(c t_m) in
   ((is_cons head_c) ∧
    let instr' = memory t_m(car_bits head_c)
    and next_c = cdr_bits head_c in
    ((is_number instr') ∧
     let instr = atom_bits instr' in
     (((instr = ^LD_instr28) ∧
       (is_cons(memory t_m next_c)) ∧
       let arg_cons_c = memory t_m(car_bits(memory t_m next_c)) in
       ((is_cons arg_cons_c) ∧
        let m_cell = memory t_m(car_bits arg_cons_c)
        and n_cell = memory t_m(cdr_bits arg_cons_c) in
        ((is_number m_cell) ∧ (is_number n_cell) ∧
         let m = (iVal(Bits28(atom_bits m_cell)))
         and n = (iVal(Bits28(atom_bits n_cell))) in
         ((¬NEG m) ∧ (¬NEG n) ∧
          (!m'.(m'<=(pos_num_of m)) ==>
               (is_cons(nth m'((memory t_m) o cdr_bits)
                           (memory t_m(e  t_m))))) ∧
          (!n'.(n'<=(pos_num_of n)) ==>
               (is_cons(nth n'((memory t_m) o cdr_bits)
                           (memory t_m (car_bits
                                          (nth(pos_num_of m)
                                            ((memory t_m) o cdr_bits)
                                            (memory t_m(e t_m)))))))))))))
      ) ∨
      ((instr = ^LDC_instr28) ∧
       (is_cons(memory t_m next_c)) ∧
       (is_atom(memory t_m(car_bits(memory t_m next_c))))
      ) ∨
      ((instr = ^LDF_instr28) ∧
       (is_cons(memory t_m next_c))
      ) ∨
      (((instr = ^AP_instr28) ∨
        ((instr = ^RAP_instr28) ∧
         (is_cons(memory t_m(e t_m))) ∧
         (car_bits(memory t_m(e t_m)) = NIL_addr) ∧
         (e t_m = cdr_bits(memory t_m
                   (car_bits(memory t_m(s t_m))))))) ∧
        (is_cons(memory t_m(s t_m)) ∧
        (is_cons(memory t_m(car_bits(memory t_m(s t_m))))) ∧
        (is_cons(memory t_m(cdr_bits(memory t_m(s t_m)))))
      ) ∨
```

```
((instr = ^RTN_instr28) ∧
 (is_cons(memory t_m(d t_m))) ∧
 (is_cons(memory t_m(cdr_bits(memory t_m(d t_m))))) ∧
 (is_cons(memory t_m(cdr_bits(memory t_m
                    (cdr_bits(memory t_m(d t_m))))))) ∧
 (is_cons(memory t_m(s t_m))) ∧
 (next_c = NIL_addr)
) ∨
(instr = ^DUM_instr28)
  ∨
((instr = ^SEL_instr28) ∧
 (is_cons(memory t_m(s t_m))) ∧
 (is_symbol(memory t_m(car_bits(memory t_m(s t_m))))) ∧
 (is_cons(memory t_m next_c)) ∧
 (is_cons(memory t_m(cdr_bits(memory t_m next_c))))
) ∨
((instr = ^JOIN_instr28) ∧
 (is_cons(memory t_m(d t_m))) ∧ (next_c = NIL_addr)
) ∨
(((instr = ^CAR_instr28) ∨ (instr = ^CDR_instr28)) ∧
 (is_cons(memory t_m(s t_m))) ∧
 (is_cons(memory t_m(car_bits(memory t_m(s t_m)))))
) ∨
((instr = ^ATOM_instr28) ∧ (is_cons(memory t_m(s t_m)))
) ∨
((instr = ^CONS_instr28) ∧
 (is_cons(memory t_m(s t_m))) ∧
 (is_cons(memory t_m(cdr_bits(memory t_m(s t_m)))))
) ∨
((instr = ^EQ_instr28) ∧
 (is_cons(memory t_m(s t_m))) ∧
 (is_cons(memory t_m(cdr_bits(memory t_m(s t_m))))) ∧
 ((is_atom(memory t_m(car_bits(memory t_m(s t_m))))) ∨
  (is_atom(memory t_m(car_bits(memory t_m
                    (cdr_bits(memory t_m(s t_m))))))))
) ∨
(((instr = ^ADD_instr28) ∨ (instr = ^SUB_instr28) ∨
  (instr = ^LEQ_instr28)) ∧
 (is_cons(memory t_m(s t_m))) ∧
 (is_cons(memory t_m(cdr_bits(memory t_m(s t_m))))) ∧
 (is_number(memory t_m(car_bits(memory t_m(s t_m))))) ∧
 (is_number(memory t_m(car_bits(memory t_m
                    (cdr_bits(memory t_m(s t_m)))))))
) ∨
((instr = ^STOP_instr28) ∧
 (is_cons(memory t_m(s t_m))) ∧ (next_c = NIL_addr))))))
```

Figure 5.1: valid_program_constraint definition

```
reserved_words_constraint mpc (memory:num->(word14->word32)) =
!tm.(state_abs (mpc tm) = top_of_cycle) ==>
  ((memory tm NIL_addr =
            bus32_symb_append #00000000000000000000000000000) ∧
   (memory tm T_addr    =
            bus32_symb_append #00000000000000000000000000001) ∧
   (memory tm F_addr    =
            bus32_symb_append #00000000000000000000000000010))
```

well_formed_free_list Informally, the free list must be a linear *cdr*
linked list, containing only *cons* cells, aside from the last cell
which is NIL. Only cells unused in any data structure that is
part of the computation (*i.e.* cells accessible from the *s, e,
c,* or *d* registers), and cells that are not reserved words (NIL
excepted) may appear in the free list. For the formal specification, there must be enough cells in the free list for the maximum
number of *cons* operations in any SECD instruction: the AP
instruction performs the operation four times. The definition
of this constraint makes use of a path function which traces
through an S-expression data structure in a memory, taking the
car or *cdr* at each step according to whether the head of the
:(bool)list argument is F or T respectively. The predicate
all_cdr_path holds when every element in the :(bool)list
argument will cause the *cdr* direction to be selected.

```
(path (mem:word14->word32) (v:word14) [] = v) ∧
  (path mem v (CONS 1 L) =
  (is_cons (mem v))  => 1 => (path mem(cdr_bits(mem v))L)
                             | (path mem(car_bits(mem v))L)
                     | v)

(all_cdr_path ([]:(bool)list) = T) ∧
(all_cdr_path (CONS h tl) = h ∧ (all_cdr_path tl))
```

Using these two functions, the desired properties of the free list are defined. The linear_free_list function requires that if any two different path arguments return the same address, then that address is the address of the NIL reserved word in memory. The not_in_free_list predicate holds when a particular address does not appear in the free list. This will apply to the high address reserved word, which is used to hold a pointer to the computation result upon executing the STOP instruction. The nonintersecting predicate states that no path in the free list leads to the same address as a path from a given cell. Lastly, the predicate n_cells_in_free_list requires that the first n addresses following *cdr's* from the *free* address all point to *cons* cells in memory. Recall that the function nth applies its 2nd argument n times to the last argument (*i.e. nth n f b = $f^n b$*).

```
linear_free_list (mem:word14->word32) (free:word14) =
  !l1 l2. (all_cdr_path l1) ∧ (all_cdr_path l2) ==>
          ¬(l1 = l2) ==>
          (path mem free l1 = path mem free l2) ==>
          (path mem free l1 = NIL_addr)

not_in_free_list (mem:word14->word32) (free:word14) (v:word14) =
  !l. (all_cdr_path l) ==> ¬(path mem free l2 = v)

nonintersecting (mem:word14->word32) (free:word14) (v:word14) =
  !l1 l2. (all_cdr_path l2) ==>
          ¬(path mem free l2 = NIL_addr) ==>
          ¬(path mem v l1 = path mem free l2)

n_cells_in_free_list (mem:word14->word32)(free:word14) (n:num) =
  !n'. (n' < n) ==> (is_cons (nth n' (mem o cdr_bits)(mem free)))
```

The full constraint is a conjunction of the above predicates. From this and the reserved_words_constraint, theorems are derived which assert that the first four addresses in the free list are not NIL_addr, and that these same addresses are all distinct.

```
well_formed_free_list memory mpc free s e c d =
 !t_m. (state_abs (mpc t_m) = top_of_cycle) ==>
         (n_cells_in_free_list (memory t_m) (free t_m) 4) ∧
         (linear_free_list (memory t_m) (free t_m))        ∧
         (let nonintersecting_with_free_list =
              (nonintersecting (memory t_m) (free t_m))
          in
          (nonintersecting_with_free_list (s t_m) ∧
           nonintersecting_with_free_list (e t_m) ∧
           nonintersecting_with_free_list (c t_m) ∧
           nonintersecting_with_free_list (d t_m)))        ∧
         (not_in_free_list (memory t_m) (free t_m) NUM_addr)
```

Garbage bits constraint To prove the correspondence of the two levels of specification for the SEL and EQ instructions, it is necessary that the bits used by the garbage collector in the subject memory locations are the same. The abstraction from the implementation to abstract memory types discards these bits, and hence the equivalence of two atoms at the abstract level is not the same as the equivalence of the representing records in memory, since the latter case includes the equality of the gc bits. For the SEL instruction, it is necessary to show that the condition for the branch in the microcode corresponded to the condition in the top level specification. The latter condition is the equivalence of the item on top of the S stack to the defined atom for T: ⊢T_atom:atom = Symb 1, while the machine does a bit by bit comparison for equality of the record on top of the stack, and the contents of memory at the reserved address T_addr = #00000000000001.

```
Garbage_bits_constraint memory mpc =
 !t_m. (state_abs(mpc t_m) = top_of_cycle) ==>
         (!x. garbage_bits(memory t_m x) = #00)
```

DEC28 assumptions The first assumption states that the :num
equivalent obtained by decrementing a 28-bit value that represents a positive integer is the same value obtained by taking
the predecessor of the :num equivalent represented by the original value. The second states that if the 28-bit value represents
a positive integer, the result of applying DEC28 will produce a
value that represents a nonnegative integer. Both of these properties are what is expected of a decrement operation, and must
be assured by the lower level implementation. The properties
enable us to assure termination of LD instruction sequences retrieving values from the environment. Both these assumptions
were discharged late in the proof.

```
DEC28_assum1 =
 !w28. (POS(iVal(Bits28 w28))) ==>
       (PRE(pos_num_of(iVal(Bits28 w28))) =
           pos_num_of(iVal(Bits28((atom_bits o DEC28) w28))))

DEC28_assum2 =
 !w28. (POS(iVal(Bits28 w28)))==>
       ¬(NEG(iVal(Bits28((atom_bits o DEC28)w28))))
```

It should be observed that each of the constraints act upon the RT
level of description. This underlines the fact that abstractions often
lose information. The simplest example is the Clock_constraint,
which applies to many points of time to which the coarse grain has
no unique corresponding points. Others of the constraints are needed
to assure the validity of the data abstractions.

5.2   Structure of the Proof

The goal for the proof of correctness is shown in Figure 5.2. Most of
the signals are simply abstracted from the medium to the coarse time
granularity. Additionally, the memory and state are abstracted to the
appropriate data type. The temporal abstraction of the single input
signal argument to the top level specification (*i.e.* the button_pin)
should be carefully considered. It can be shown that this input affects

the next state of the machine at the medium time granularity in only
three places in the microcode, and these three places correspond to
points in coarse grain time, since they are all *major states*.

```
((clock-constraint SYS_Clocked)                          ∧
 (reserved_words_constraint mpc memory)                  ∧
 (well_formed_free_list memory mpc free s e c d) ∧
 Garbage_bits_constraint memory mpc                      ∧
 (valid_program_constraint memory mpc button-pin s e c d))
 ==>
 (SYS memory SYS_Clocked
        mpc s0  s1  s2  s3
        button_pin
        flag0_pin   flag1_pin  write_bit_pin  rmem_pin
        bus_pins    mar_pins
        s   e   c   d   free
        x1  x2  y1  y2  car  root  parent  buf1  buf2  arg)
 ==>
 SYS_spec  ((mem_abs o memory)   when (is_major_state mpc))
           (s                    when (is_major_state mpc))
           (e                    when (is_major_state mpc))
           (c                    when (is_major_state mpc))
           (d                    when (is_major_state mpc))
           (free                 when (is_major_state mpc))
           (button_pin           when (is_major_state mpc))
           ((state_abs o mpc)    when (is_major_state mpc))
```

Figure 5.2: RTL ⊃ top level goal

Although the lower (RT) level definition of the system for this
proof consists of the wiring together of high-level components defined
behaviourally, it is not feasible to undertake the proof of the top level
goal directly. The size of terms generated alone makes this impossible
to manage. The proof is instead undertaken in several stages.

First, a simplified, flattened specification for the RT level definition is obtained. Second, this specification is used to derive theorems for the change of state effected by each microcode instruction. These theorems represent the cumulative effect of asynchronous
events through the phases of one full clock cycle, and roughly cor-

respond to the "Phase level" of description of Anceau [1]. Third, a sort of microcode simulation uses the microinstruction theorems to step through the microcode to produce theorems that summarise the behaviour of instruction sequences. This corresponds to the "Microprogramming level" of Anceau. Fourth, the sequence theorems are used to prove a liveness property for the system. And last, the state transitions resulting from the register transfer level definition are proved to correspond to those defined at the top level, for all transitions possible under the constraints.

As with the specification of the system in the previous chapter, the size and number of theorems prevents inclusion in their entirety. Using the HOL pretty printer, a complete listing of the theorems is estimated at over 15,000 lines (250 pages!), three times as much as the definitions. A description of each stage of the proof follows.

5.3   Unfolding the System Definition

The register transfer level of description consists of a conjunction of behavioural descriptions of major components. For example, the control unit consists of a *state register* (with five fields), a *ROM* and a *decode* unit. A specification for the conjunction of these three subcomponents is not appreciably simpler than the conjunction of the component behaviours individually. It is only when the whole system is assembled that simplifications become possible. A good example involves the **one_asserted** property of the ALU control lines. The property is a constraining part of the definition of the ALU, but it is provable from the control unit definition that the ALU control lines have this property. This constraint can be eliminated when the control unit and datapath are conjoined.

Thus we abandon the simple hierarchical approach, wherein both an implementation and specification are defined independently for each component of the hierarchy and a correctness proof of the top component is achieved by using the correctness results of subcomponent parts. Instead starting with the definitions of the major subcomponents of the chip, each is simplified by unfolding all definitions, and UNWIND'ing[1] existentially quantified variables (hidden wires) where possible. When the only remaining occurrence of the existentially quantified variable is the left side of a single equation, it may be PRUNE'd, or eliminated from the expression. This is continued up to the top of the hierarchy where more substantial simplifications can be made. The process of creating these proofs is relatively straightforward, complicated only by the relatively large size of terms involved. However, the large term size forces careful management of the proof process. Often only primitive rules and tactics are usable, as the more powerful tactics readily exhaust memory.

First we undertake unfolding the datapath and control unit definitions. A tactical proof approach is taken, demanding careful goal preparation when large terms are involved. The ease of tactical proof methods compared to forward proof when used to "massage" a term to a particular form make the investment in time worthwhile. Often the desired terms can be derived (or almost derived) in the first instance by installing a component as a goal and unfolding subcomponent definitions by rewriting.

The goal for the datapath introduces a pair of antecedents: the one_asserted property applied to the ALU control lines, and the clock_constraint, while the consequent is the equality of the DP and its unfolded and simplified version. The proof unfolds all definitions and let bindings, rewrites with the two antecedents to simplify the resulting expressions, flattens and reorders all conjuncts, and moves the universally quantified time parameter outwards to enclose all conjuncts. The single existentially quantified value alu remains, outside the level of (and enclosing) the quantified time parameter in the theorem. This variable cannot be eliminated unless a particular alu operation is specified.

The control unit is treated similarly. The goal is a simple equality of the CU expression and its unfolded and simplified version: a conjunction of equations for each state variable and output. All definitions are unfolded and let expressions unfolded, the expression for the next state values of the five field state register is split

---

[1] UNWIND is a HOL conversion which unfolds the equations for existentially quantified variables that occur as conjuncts in an expression.

into distinct expressions for each field, the time parameter is moved out to enclose all conjuncts, the seven existentially quantified values: rom_out, nextmpc, next_s0, next_s1, next_s2, next_s3, and push_or_pop are unwound and pruned, and a few logical simplifications are made. The simplified form of the control unit definition is used to derive a theorem showing that the ALU control outputs it generates have a one_asserted property.

In later stages of the proof, it will be necessary to evaluate each control unit output value. The equations are typified by the add output:

(add t_m= (Alu_field(ROM_fun(mpc t_m)) = #0010)).

Given a fixed value for mpc t_m, we can readily evaluate the subterm Alu_field(ROM_fun(mpc t_m)) to a :word5 constant. The intensive computation required to prove inequality of two constants of a specified $word_n$ type makes it desirable to prove an exhaustive set of theorems for all possible values of each subfield of the microcode ROM output. Rather than repeatedly proving the cases for each of the 400 ROM addresses, this requires 68 theorems[2] all told. These theorems are of the form:

```
Alu_base_0010 =
⊢ (Alu_field(ROM_fun(mpc tₘ)) = #0010) ==>
  ((Alu_field(ROM_fun(mpc tₘ)) = #0001) = F) ∧
  ((Alu_field(ROM_fun(mpc tₘ)) = #0010) = T) ∧
  ...
  ((Alu_field(ROM_fun(mpc tₘ)) = #1011) = F) ∧
  ((Alu_field(ROM_fun(mpc tₘ)) = #1100) = F)
```

Note that these theorems are designed to be used for substitution rather than the less efficient rewriting.

In addition, theorems for the value of the Inc9 function, used to calculate the next mpc address for sequential microcode, for each of 400 addresses are proved, typified by the theorem for the lowest address.

---

[2] A theorem for each possible value is required, including the no-op value, hence there are 13 possible values for the alu and test fields, 18 for the write field, and 24 for the read field.

```
⊢ Inc9 #000000000 = #000000001
```

At the next level, SECD is simplified by unfolding the PF and DP
components, but not the CU. The one_asserted constraint is eliminated, the clocked_constraint is included, hidden lines connected
directly to input or output pads are unwound and pruned, including button, flag0, flag1, write_bit, and mar_bits, and all remaining existentially quantified variables are moved to the outermost
level.

At the top of the definition hierarchy, the static RAM definition of memory is added in, and the CU and SECD simplifications
are applied to create one flattened expression for the SYS implementation. All internal read and write control lines are unwound
and pruned. The only remaining existentially quantified variables,
bus_bits, mem_bits, and alu, are quantified over all conjuncts. The
values of these variables depend on the control unit outputs, which
in turn depend on the microinstruction being executed. The most
important step involves moving the time parameter outside the existentially quantified internal lines, and replacing these time-varying
lines with static values. The validity of this transformation is expressed in the following theorem, which states that the existence of
a function from time to values for which some property holds at all
times is equivalent to saying that at all times, there exists a value
for which the property holds.

```
⊢ (?(a:num->*). !(t:num). P t(a t)) =
  (!(t:num).    ?(a_t:*). P t a_t)
```

A set of specialised conversions are used to apply the theorem to precise locations in the goal term, resulting in the theorem of Figure 5.3.
Importantly, the time parameter has been specialised to an arbitrary
value t_m, so that the expression can be evaluated when constraining some value at this time; in fact the simplified expression for the
system description is evaluated repeatedly under the constraint mpc
t_m= x, where x is one of the 399 microcode addresses. From this
stage onward, equivalence is no longer attempted, so the final form

```
[clock_constraint SYS_Clocked; ^SYS_imp]
⊢ ? bus_bits_t mem_bits_t alu_t.
    (mpc(SUC t_m) =
        ((((Test_field(ROM_fun(mpc t_m)) = #0001) ∨
          (Test_field(ROM_fun(mpc t_m)) = #0011) ∧ field_bit bus_bits_t ∨
          (Test_field(ROM_fun(mpc t_m)) = #0100) ∧ mark_bit bus_bits_t ∨
          (Test_field(ROM_fun(mpc t_m)) = #0101) ∧ (bus_bits_t = arg t_m) ∨
          (Test_field(ROM_fun(mpc t_m)) = #0110) ∧
          LEQ_prim(arg t_m)bus_bits_t ∨
          (Test_field(ROM_fun(mpc t_m)) = #0111) ∧
          (cdr_bits bus_bits_t = NIL_addr) ∨
          (Test_field(ROM_fun(mpc t_m)) = #1000) ∧ is_atom bus_bits_t ∨
          (Test_field(ROM_fun(mpc t_m)) = #1001) ∧
          (atom_bits bus_bits_t = ZERO28) ∨
          (Test_field(ROM_fun(mpc t_m)) = #1010) ∧ button_pin t_m ∨
          (Test_field(ROM_fun(mpc t_m)) = #1011))
        => A_field(ROM_fun(mpc t_m))
        | ((Test_field(ROM_fun(mpc t_m)) = #1100)
        => s0 t_m
        | ((Test_field(ROM_fun(mpc t_m)) = #0010)
        => Opcode arg t_m | Inc9(mpc t_m)))))                    ∧
    (s0(SUC t) =
        ((Test_field(ROM_fun(mpc t_m)) = #1011)
        => Inc9(mpc t_m)
        | ((Test_field(ROM_fun(mpc t_m)) = #1100) => s1 t_m | s0 t_m))) ∧
        ...
    (memory(SUC t_m) =
        ((Write_field(ROM_fun(mpc t_m)) = #00001)
        => Store14(mar_pins t_m)(bus_pins t_m)(memory t_m) | memory t_m)) ∧
        ((Read_field(ROM_fun(mpc t_m)) = #00010) ==>
          (bus_bits_t = mem_bits_t))                             ∧
        ...
    (rmem_pin t_m = (Read_field(ROM_fun(mpc t_m)) = #00010)) ∧
        ((Alu_field(ROM_fun(mpc t_m)) = #0001) ==>
          (alu_t = DEC28(atom_bits(arg t_m))))                  ∧
        ...
    (((¬(Write_field(ROM_fun(mpc t_m)) = #00001))
        => (mem_bits_t = bus_pins t_m) | (bus_pins t_m = bus_bits_t)) ∧
        (¬(Write_field(ROM_fun(mpc t_m)) = #00001)∧
        (Read_field(ROM_fun(mpc t_m)) = #00010) ==>
          (bus_pins t_m = memory t_m(mar_pins t_m)))            ∧
        ...
    (s t_m = ((Write_field(ROM_fun(mpc t_m)) = #00110)
              => cdr_bits bus_bits_t | s(PRE t_m)))             ∧
        ...
    (flag1_pin t_m = (mpc t_m = #000101011) ∨ (mpc t_m = #000011000))
```

Figure 5.3: Base_thm: the RTL definition simplified

of the theorem has the clock_constraint and the SYS definition[3] as assumptions, and the conclusion is the simplified expression for the system, with the three existentially quantified variables. Also the conjunct giving the initial value of the mpc (*i.e.* at time $t_m=$ 0) is dropped from the expression. Representative samples of each type of conjunct in the resulting theorem are shown in Figure 5.3. Generating this stage of the proof requires over 74,000 primitive inferences[4], with nearly half this number devoted to the proofs of the 68 theorems for inequality of the microcode ROM output fields.

There is one final observation to make of the theorem of Figure 5.3. The expressions for control unit state values (*i.e.* mpc, s0, ..., s3) and the memory all define their values at time SUC t_m as a function of values at time t_m, while the expressions for data path state values (*i.e.* s, e, c, d, free, x1, ...) give their values at time t_m as a function of values at time PRE t_m. The reason for the difference between the registers is tied to the temporal abstraction from the finer to medium grains of time, and the different design and clocking of the control unit and datapath registers, illustrated in Figures 4.6 and 4.7. Moreover, we require the arg register to pass through an opcode value to the control unit as it is being read from the bus, and all datapath registers share a consistent definition. It also seems quite natural: the control unit drives the datapath which can cause a change in register contents now, and it also prepares its next mpc value for the next instruction. The reason the memory matches the control unit rather than datapath is an arbitrary choice, but made possible by the fact that there cannot be simultaneous read and write operations on memory, and thus a value written at some time t_m cannot be accessed until time SUC t_m. Hence, the representation is consistent with component behaviour.

---

[3]This constraint will be abbreviated as SYS_imp henceforth.

[4]It may interest some readers to learn that this figure was over 440,000 when the proof was performed using the earlier HOL88 Version 1.11. The improvement is believed to be entirely due to refinements within the HOL system, as there were no significant changes to the proofs involved.

5.4   Phase Stage: Effect of Each Microinstruction

The simplified expression for the system is next utilized to prove theorems giving the effect of each microinstruction execution on the system state. A sample theorem for the system when the value #001100001 is in the MPC9 register is shown in figure 5.4. This is the first instruction of the microcode sequence for the LDF instruction, and effects a transfer of the content of the E_reg register to the X2_reg register.

The large number of such theorems (337 excluding garbage collection sequences) requires that they be generated without direct user intervention. A single proof function is designed for this purpose, capable of generating the required theorems in a forward proof manner, without prior statement of the specific content of each. Such a proof function is developed by working through specific sample theorems. As failures or unsatisfactory theorems are observed, the proof function is revised. Generating the set of theorems takes many hours[5], and requires in excess of one million primitive inferences. The theorems are divided among seven theories, each containing adjacent instruction code sequences, to reduce search times for theorems at the next proof stage.

The phase_proof_fcn is applied to a :word9 (microcode) address. It proceeds as follows.

1. Fetch the theorem defining the value of the microcode function ROM_fun for the given address, add the assumption that (mpc t_m) is equal to the address, and substitute this term for the address in the theorem. The resulting theorem has the form:
   . ⊢ ROM_fun(mpc t_m) = #00000000000000000110001000
   In this case and similarly in all following theorems, the dot to the left of the turnstile symbol represents the assumption (mpc t_m = #001100001).

---

[5] In the original proof development using HOL88 1.9 built with Franz Lisp, each attempt took up to 36 hours running on a dedicated Sun 3/60 workstation with 16 megabyte memory. Currently, this is estimated to take 12 hours on a 16 megabyte Sun Sparc 2, using HOL88 2.0 built with Allegro Common Lisp.

```
[clock_constraint SYS_Clocked; ^SYS_imp]
⊢ (mpc t_m = #001100001) ==>
  (mpc(SUC t_m) = #001100010)            ∧
  (s0(SUC t_m) = s0 t_m)                 ∧
  (s1(SUC t_m) = s1 t_m)                 ∧
  (s2(SUC t_m) = s2 t_m)                 ∧
  (s3(SUC t_m) = s3 t_m)                 ∧
  (memory(SUC t_m) = memory t_m)         ∧
  (x2 t_m = e(PRE t_m))                  ∧
  (rmem_pin t_m = F)                     ∧
  (buf1 t_m = buf1(PRE t_m))             ∧
  (buf2 t_m = buf2(PRE t_m))             ∧
  (mar_pins t_m = mar_pins(PRE t_m))     ∧
  (s t_m = s(PRE t_m))                   ∧
  (e t_m = e(PRE t_m))                   ∧
  (c t_m = c(PRE t_m))                   ∧
  (d t_m = d(PRE t_m))                   ∧
  (free t_m = free(PRE t_m))             ∧
  (x1 t_m = x1(PRE t_m))                 ∧
  (car t_m = car(PRE t_m))               ∧
  (arg t_m = arg(PRE t_m))               ∧
  (parent t_m = parent(PRE t_m))         ∧
  (root t_m = root(PRE t_m))             ∧
  (y1 t_m = y1(PRE t_m))                 ∧
  (y2 t_m = y2(PRE t_m))                 ∧
  (write_bit_pin t_m = T)                ∧
  (flag0_pin t_m = F)                    ∧
  (flag1_pin t_m = F)
```

Figure 5.4: Theorem for execution of microcode instruction at address 97

2. Apply the wordn_CONV conversion to transform the 27 bit constant value output to the equivalent form expressed as the application of Word27 to a 27 bit *(bool)bus*, in this case:
   Word27(Bus F...(Bus T(Bus F(Bus F(Wire F)))))...).
   For each field of the ROM output, apply the field selector function, unfold its definition, and convert the resulting *bus* form of the field value back into the *word_n* constant representation.

Five theorems, one for each field, are produced, typically of the form:

. ⊢ Read_field(ROM_fun(mpc t_m)) = #01000
. ⊢ Write_field(ROM_fun(mpc t_m)) = #01100
. ⊢ Alu_field(ROM_fun(mpc t_m)) = #0000
. ⊢ Test_field(ROM_fun(mpc t_m)) = #0000
. ⊢ A_field(ROM_fun(mpc t_m)) = #000000000.

3. The four theorems for the Test_field, Alu_field, Read_field, and Write_field values are then each resolved with the control unit theorems for the appropriate constants and fields, and split into conjuncts, giving a set of theorems each having the form:

. ⊢ (Read_field(ROM_fun(mpc t_m)) = #00001) = F, or
. ⊢ (Read_field(ROM_fun(mpc t_m)) = #01000) = T.

4. The relevant Inc9 theorem is retrieved from Inc9_proofs and specialised:

. ⊢ Inc9(mpc t_m) = #001100010.

5. The expressions for state flag values are evaluated after substituting the mpc value to generate a theorem of the form:

. ⊢ (mpc t_m=#000010110) ∨ (mpc t_m=#000011000) = F.

6. The set of all the preceding theorems is assembled as a list, and substituted into the Base_thm from the previous stage, using the primitive inference rule SUBST and a template. A series of rewrites reduces constant boolean expressions and removes eliminated existentially quantified variables, and is followed by a PRUNE'ing of remaining existentially quantified variables.

7. Any remaining existentially quantified variables will occur in a conjunct in which only one field is defined. This occurs when a 14 bit value is transferred over the bus, for example. These are eliminated by a specialised rule that uses theorems about the existence of fields of an existentially quantified variable:

car_bits_thm = ⊢ !y. ?x. car_bits x = y
cdr_bits_thm = ⊢ !y. ?x. cdr_bits x = y.

8. Discharge the original assumption of the value of the mpc.

Only one of the 337 microinstruction theorems requires additional simplification. The last microinstruction (308) of the sequence for the STOP instruction stores a pointer to the result of the computation in the *cdr* field of the highest memory address. This is effected by reading the S_reg register and writing to memory. Only the *cdr* field of the bus is driven, so only that field of the memory record has a determinable value, and even the record type is indeterminate. This is clearly an example of imprecise specification that formal methods will bring to our attention. It is a moot point whether this could, in practice, present any difficulty in the system operation. However, the problem here is the inability of the proof function to eliminate the existentially quantified value bus_bits_t, leaving the subterm:

```
. ⊢ ...
   (?bus_bits_t.
   (memory(SUC t_m) =
   Store14(mar_pins(PRE t_m))bus_bits_t(memory t_m)) ∧
   (cdr_bits bus_bits_t = s(PRE t_m)) ∧
   ...}
```

To resolve this problem, the expression for the state of memory was changed to:

```
(!a.((a = mar_pins(PRE t_m))
   => (cdr_bits(memory(SUC t_m)a) = s(PRE t_m))
   | (memory(SUC t_m)a = memory t_m a)))
```

This expression only defines part of the memory output at the subject address, and allows the expression for bus_bits_t to be substituted in the memory expression, leaving a single occurrence of bus_bits_t in the theorem, which can be pruned.

5.5 Microprogramming Stage: Symbolic Execution

Each one of the 337 lemmas about the RTL description gives the change in state over one unit of time. These results are next combined to give the state change over a sequence of steps.

The top level system definition SYS_spec describes a four state finite state machine, with two transitions from each of three states controlled by the button input, an initial transition from a deterministic startup state, and 18 possible transitions, one for each machine instruction code, from the fourth state. Further, four instruction sequences have a branch conditional upon some function of the state, and one (the sequence for the LD instruction) contains two loops. Thus there are a minimum of 29 paths in the microcode to examine. Additionally, several instruction sequences call subroutines, and subroutine calls are nested. Under the well_formed_free_list constraint, the subroutines have a deterministic execution time as garbage collection never occurs.

The proofs generated in this section require a total of 430,000 primitive inferences. The number of inferences for each proof is in quite direct proportion to the number of microcode instructions in the sequence.

This section begins with a simple proof for the initial transition, and continues with a description of the general approach to developing the longer sequence proofs, using the LDF instruction as the representative example. Following this is a description of the more complicated LD sequence proof, and the application of the developed methodology to the proofs for the most complex AP and RAP instruction sequences.

5.5.1 The initial transition

We begin by considering the transition from the startup state to the first *major state* which the system reaches. The first proof stage has produced an unfolded definition of the system which includes a conjunct giving the initial mpc value. Taking this first conjunct gives the theorem:

.. $\vdash$ mpc 0_m= #000000000[6].

Applying *modus ponens* to the phase stage theorem for microinstruction 0, SYS_lemma_0, and the previous theorem, and taking the first

---

[6]The assumptions are the clock constraint and SYS_imp.

conjunct gives the theorem:

.. ⊢ mpc(SUC 0_m) = #000010110.

Applying the predicate *is_major_state* to *mpc* at both time 0_m and SUC 0_m, then unfolding all definitions produces the two theorems:

.. ⊢ ¬is_major_state mpc 0_m

.. ⊢ is_major_state mpc(SUC 0_m).

The next theorem appears deceptively simple:

.. ⊢ TimeOf(is_major_state mpc)0_m = SUC 0_m.

The proof begins by unfolding the definition of TimeOf, but this introduces the SELECT operator @ used in its definition. It is necessary to show that there is a unique value which satisfies the body of the operator, and a specially devised tactic *SELECT_UNIQUE_TAC* [24] does just that, splitting the goal into two parts. The first requires a proof that the value SUC 0_m satisfies the predicate, while the second must show that the satisfying value is unique. This property is provable from the definition of IsTimeOf, and is captured by the theorem IsTimeOf_IDENTITY:

⊢ !n f t1 t2.

    IsTimeOf n f t1 ∧ IsTimeOf n f t2 ==> (t1 = t2)

The final theorem discharges the assumptions, and uses the same form of abstraction as appears in the top level goal of Figure 5.2.

```
⊢ clock_constraint SYS_Clocked ==>
  ^SYS_imp ==>
  (((state_abs o mpc) when (is_major_state mpc)) 0_c = idle)
```

5.5.2   The general approach: LDF

The proof for the initial transition is simply concerned with the time the machine gets to a *major state*, and what that state is. For most of the remaining transitions, the state of the rest of the system, comprising the external memory as well as the on-chip registers, is of equal concern.

Central to the method is the idea of symbolic execution. Given an initial theorem expressing the accumulated change to the system state after execution of some microcode sequence, one conjunct gives the value of mpc at the next point in time. The next microinstruction is symbolically executed by applying MATCH_MP (a HOL variant of Modus Ponens) to the appropriate microinstruction phase theorem and the mpc conjunct from the initial theorem. The resulting theorem is rewritten with the initial theorem conjuncts to incorporate the previously accumulated computation. It is also necessary (and convenient) to prove whether or not the system is in a *major state* at each step. This second result is accumulated in a theorem giving a defined interval in which the system is not in a *major state*. The sequence proof ends when a *major state* is reached, whereupon a specific value for the **Next** time the system is in a *major state* is determined.

Once again, the number of theorems demands minimising user intervention. Another proof function is designed to effect the single step computation described above, producing a new pair of theorems from a theorem pair argument. A higher order function calls it recursively to effect a series of steps, producing a pair of theorems summarising the computation of a sequence of microinstructions. The result of executing a machine instruction, in this case the LDF instruction, is captured by the two theorems in Figure 5.5.

The mpc value at the end of the sequence corresponds to the top_of_cycle state. The **memory** has had two cells altered: the first cell has pointers to the code argument to the LDF instruction, and the current environment **e**, effectively representing a closure. The second cell has pointers to the first cell, and the original **s** value. The **s** pointer is updated to point to the latter cell, effectively storing the closure on top of the S stack. The **c** pointer now points to the rest of the control list following the LDF instruction and its argument. Notice that the **cdr_bits** operation is performed on different memories: for the **s** value on the original memory, and for the **c** value after the memory has been updated with the two rewritten cells. The **free** pointer is similarly updated to the the third cell in the original free list. The content of the other registers is irrelevant to the abstracted state, and they are removed in the final theorem shown. The expression for the updated memories appears several times, so they have been bound in **let** expressions to improve readability by reducing the term size.

```
LDF_state =
...... ⊢ let mem1 =
            Store14(free t_m)
                    (bus32_cons_append
                    #00 RT_CONS
                    (car_bits(memory t_m
                       (cdr_bits(memory t_m(c t_m)))))
                    (e t_m))
                    (memory t_m)
        in
        let mem2 =
            Store14(cdr_bits(memory t_m(free t_m)))
                    (bus32_cons_append #00 RT_CONS
                                       (free t_m)(s t_m))
                    mem1
        in
        ((mpc(t_m+26) = #000101011) ∧
         (memory(t_m+26) = mem2) ∧
         (s(t_m+26) = cdr_bits(memory t_m(free t_m))) ∧
         (e(t_m+26) = e t_m) ∧
         (c(t_m+26) = cdr_bits(mem2(cdr_bits(mem2(c t_m))))) ∧
         (d(t_m+26) = d t_m) ∧
         (free(t_m+26) =
          cdr_bits(mem1(cdr_bits(memory t_m(free t_m)))))))

LDF_Next = ...... ⊢ Next t_m(t_m+26)(is_major_state mpc)

The assumptions for both theorems are:
[ clock_constraint SYS_Clocked
; ^SYS_imp
; reserved_words_constraint mpc memory
; well_formed_free_list memory mpc free s e c d
; mpc t_m = #000101011
; opcode_bits(memory t_m(car_bits(memory t_m(c t_m)))) = #000000011
]
```

Figure 5.5: Microprogramming stage theorems for LDF instruction

The first sequences proved are the microcode subroutines. The proof commences with initial theorems of the form (for the *consx1x2* subroutine):

```
.. ⊢ (mpc t_m=#101000101) ∧ ((free(PRE t_m)=NIL_addr) = F)}

⊢ !t_m". ((PRE t_m)<t_m") ∧ (t_m"<t_m)
           ==>¬is_major_state mpc t_m"
```

with the assumptions of the first identical to the conjuncts of its conclusion. The second theorem is vacuously true, since it describes the property over an empty interval.

The first theorem evolves at each step as described earlier. The second uses the theorem **Next_step**:

```
⊢ !ts tf f.(¬f tf) ∧ (!t.(ts<t) ∧ (t<tf)==>(¬f t)) ==>
          (!t.(ts<t) ∧ (t<(SUC tf)) ==> (¬f t))
```

Once the value of the mpc at the given time is used to prove the system is *not* in a *major state*, it is combined with the accumulated interval theorem and **Next_step** to give a new theorem, covering an interval one time unit longer. The results for the *consx1x2* subroutine are shown in Figure 5.6.

The proof strategy for the instruction transition sequences sees the derivation of theorems for each of the subroutines exclusive of garbage collection. Next, the sequence common to all instructions, essentially a *fetch instruction* operation starting at top_of_cycle state, is proved up to where the control stream branches to each individual instruction code sequence. For each of the 18 instructions, an assumption about the value of the next SECD instruction is added to the initial theorem, and one more step symbolically executed. The resulting theorems are used as arguments to the proof function, which continues advancing the symbolic execution through the microcode until a *major state* is encountered.

The starting theorem for the proof function in these cases is slightly different from that used for the subroutines. The final theorem must express the value for all state variables at the same time, in

```
Consx1x2_state =
.... ⊢ (mpc(SUC(SUC(SUC(SUC t_m)))) = s0 t_m) ∧
        (s0(SUC(SUC(SUC(SUC t_m)))) = s1 t_m) ∧
        (s1(SUC(SUC(SUC(SUC t_m)))) = s2 t_m) ∧
        (s2(SUC(SUC(SUC(SUC t_m)))) = s3 t_m) ∧
        (s3(SUC(SUC(SUC(SUC t_m)))) = #000000000) ∧
        (memory(SUC(SUC(SUC(SUC t_m)))) =
        Store14 (free(PRE t_m))
                (bus32_cons_append #00 RT_CONS(x1(PRE t_m))
                                               (x2(PRE t_m)))
                (memory t_m)) ∧
        (bus_pins(SUC(SUC(SUC t_m))) =
        bus32_cons_append #00 RT_CONS(x1(PRE t_m))(x2(PRE t_m))) ∧
        (rmem_pin(SUC(SUC(SUC t_m))) = F) ∧
        (buf1(SUC(SUC(SUC t_m))) = buf1(PRE t_m)) ∧
        (buf2(SUC(SUC(SUC t_m))) = buf2(PRE t_m)) ∧
        (mar_pins(SUC(SUC(SUC t_m))) = free(PRE t_m)) ∧
        (s(SUC(SUC(SUC t_m))) = s(PRE t_m)) ∧
        (e(SUC(SUC(SUC t_m))) = e(PRE t_m)) ∧
        (c(SUC(SUC(SUC t_m))) = c(PRE t_m)) ∧
        (d(SUC(SUC(SUC t_m))) = d(PRE t_m)) ∧
        (free(SUC(SUC(SUC t_m))) =
        cdr_bits(memory t_m(free(PRE t_m)))) ∧
        (x1(SUC(SUC(SUC t_m))) = x1(PRE t_m)) ∧
        (x2(SUC(SUC(SUC t_m))) = x2(PRE t_m)) ∧
        (car(SUC(SUC(SUC t_m))) = car(PRE t_m)) ∧
        (arg(SUC(SUC(SUC t_m))) = arg(PRE t_m)) ∧
        (parent(SUC(SUC(SUC t_m))) = parent(PRE t_m)) ∧
        (root(SUC(SUC(SUC t_m))) = root(PRE t_m)) ∧
        (y1(SUC(SUC(SUC t_m))) = y1(PRE t_m)) ∧
        (y2(SUC(SUC(SUC t_m))) = y2(PRE t_m)) ∧
        (write_bit_pin(SUC(SUC(SUC t_m))) = F) ∧
        (flag0_pin(SUC(SUC(SUC t_m))) = F) ∧
        (flag1_pin(SUC(SUC(SUC t_m))) = F)

Consx1x2_nonmajor =
.... |- !t'_m.(PRE t_m) < t'_m ∧ t'_m < (SUC(SUC(SUC(SUC t_m)))) ==>
              ¬is_major_state mpc t'_m

The assumptions of both theorems are:
[ clock_constraint SYS_Clocked
; ^SYS_imp
; (free(PRE t_m) = NIL_addr) = F
; mpc t_m = #101000101
.]
```

Figure 5.6: Microprogramming stage theorems for consx1x2 subroutine

terms of the initial values at time t_m, unlike the form of the microinstruction theorems typified by the example in Figure 5.4. Thus, only the conjuncts describing the next state of mpc, s0, s1, s2, s3, and memory from the microinstruction lemma for the top_of_cycle address form the initial theorem. The starting theorem for the range in which it is not in a *major state* differs as well:

$$\vdash \ !t_m". (t_m < t_m") \ \wedge \ (t_m" < (\text{SUC } t_m)) \ ==>$$
$$\neg \text{is\_major\_state mpc } t_m".$$

The sequence proof function is designed as a recursive function applied to a single step proof function. It is quite general, and is used for the proof of theorems for all the sequences, including the subroutines, the instruction fetch sequence, and the sequences for each machine instruction. The decision tree for the recursive function is shown in Figure 5.7. When the next instruction is not a

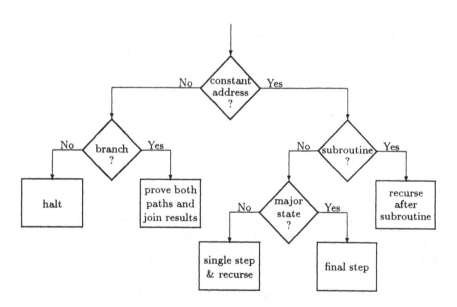

Figure 5.7: Decision Tree for Recursive Microprogramming Stage Proof Function

simple constant, it must be either a conditional branch or a case split on the SECD instruction code. In the latter case, user intervention is required to add an appropriate assumption determining which instruction code is to be executed, as well as any other assumptions related to that instruction. These assumptions are taken directly from the valid_program_constraint. In the case of a simple branch, both legs of the proof are followed, one by adding the conditional expression as an assumption, and the other by adding the negation. Subroutine calls encountered in the computation sequence use the subroutine theorems to accumulate the effect of the subroutine execution, using an interval proof function. Most instructions will cause a recursive call on the result of executing one more microinstruction. When a *major state* is encountered execution must stop, but only after the step function completes updating the state of the datapath registers. This is required since these registers are not part of the starting state theorem, and their final value needs to catch up one step. The state values that are not relevant to the next stage of the proof are culled out at this point. At the last step, the time parameters are changed to a sum of t_m and a constant. A further simplification binds the intermediate values of memory in let expressions to produce a theorem pair as in Figure 5.5.

5.5.3   Proving the complex sequences

The proof function just described fails to prove three sequences. Two of the three, the sequences for executing an AP or RAP instruction, produce terms which are so large that memory faults occur. The excessive term size arises from two factors: the time parameters for the resulting state are expressed in terms of some repeated number of applications of SUC to the start time t_m, and the expression defining the state of memory after successive store operations occurs repeatedly. Both of these are reduced in the sample LDF theorem of Figure 5.5, but the reduction takes place after the theorem proof is essentially complete. As an example, the completely unfolded (*i.e.* with all let expressions removed) pretty printed version of the state theorem for the RAP instruction is 2,436 lines long, compared to 50 lines for the form containing let expressions. It is considerably more difficult, if it is indeed possible, to effect the proof without unfolding let bindings in intermediate results. The overhead of converting time parameters to simple sums of t_m and a constant is reduced by performing it only once, at the end of the proof.

The proof for the LD sequence has to demonstrate termination of execution of the microcoded "while" loops, and the accumulated effect of executing these loops is proved first. This part of the proof method provides an approach which is later successfully applied to the two sequences just described. The major difference is that segments of the microcode are proved just as for the subroutine sequences, the time parameter is converted to a sum with a constant, and then the intervals are assembled much as for the subroutines in the earlier proofs. A complex conversion from a time parameter consisting of a sum of several constants is required once only, and the proofs successfully complete without memory faults. While this is not significant in terms of logical formalism, it is typical of the problems encountered when applying formal methods to substantial systems. Much time is spent managing complexity. Creating a proof is often not in itself a problem; creating a large proof within fixed resource limits can be.

Floyd-Hoare logic [23] provides a model for reasoning about *while* loop constructs within imperative languages. The form of rules in this logic is:

⊢{*precondition*} code-segment {*postcondition*},

which asserts that when the precondition holds, if execution of code-sequence terminates, it will result in a state in which postcondition holds. The *WHILE*-rule states:

```
    ⊢ {P ∧ S} C {P}
--------------------------------
⊢ {P} WHILE S DO C {P ∧ ¬S}
```

P is an *invariant* of C whenever S holds. S is the test condition for the while loop. Executing the program segment will preserve the truth of P, and upon termination, also assure that S no longer holds (otherwise it will not terminate).

The microcode sequence implementing the while loops consist of 5 instructions, shown in Figure 5.8. The effect of executing this loop once is to replace the contents of the $x1$ register with the *cdr*

```
L("LD0");
            rarg    ;                        ; (test_zeroflag ("LD1")) ;
            rx1     ; wmar                   ; (inc ()) ;
            rmem    ; wx1                    ; (inc ()) ;
                    ; wbuf1    ; dec         ; (inc ()) ;
            rbuf1   ; warg                   ; (jump ("LD0")) ;
L("LD1");
```

Figure 5.8: The While Loop Microcode

of its initial value, and to decrement the value in the *arg* register, transferred via the **buf1** register. The *test* condition is that the the **arg** register does not contain the constant zero, otherwise a jump past the loop is effected. The *invariant* must express a relation between the contents of the **arg** and **x1** registers, in essence that the value X obtained by taking the *cdr* of **x1 arg** number of times remains constant. Upon exiting from the loop, the **arg** value will be zero, and hence the value of X will be equal to **x1**. The formal expression of these ideas requires the explicit incorporation of a time parameter, as well as incorporating the remainder of the state values, including the **mpc**.

Before proving any theorem that summarises the repeated computation of the loop, termination must be established. The initial value in **arg** must be non-negative, and this will be a constraint in the final proof. We must prove, under the two **DEC28** assumptions, that: n applications of **DEC28** to the 28 bit value that represents n gives the value **ZERO28**, and for any fewer than n applications, the value is not **ZERO28**. This is captured in the following two theorems:

```
loop_terminates_lemma =
.. ⊢ !w28.
      ¬NEG(iVal(Bits28 w28)) ==>
      (nth(pos_num_of(iVal(Bits28 w28)))(atom_bits o DEC28)w28 =
      ZERO28)

nth_DEC_NOT_ZERO =
.. ⊢ !n w28.
      ¬NEG(iVal(Bits28 w28)) ==>
      n < (pos_num_of(iVal(Bits28 w28))) ==>
      ¬(nth n(atom_bits o DEC28)w28 = ZERO28)
```

The next step is to derive a pair of theorems, shown in Figure 5.9 summarising the accumulated effect of one pass through the while loop, capturing the sense of the *loop invariant*. We assume that the termination condition is false at the start.

Next, using the nth_DEC_NOT_ZERO theorem and the loop1_ theorems, the accumulated computation effect of up to n iterations through the loop is captured in the pair of theorems in Figure 5.10. The proof proceeds by induction on n.

Finally, loop_terminates_lemma is used to derive that the exit condition hold after n iterations. This whole procedure is repeated for the second while loop. The final theorem for the LD instruction is derived by assembling theorems for both loop sequences and the initial, final, and in-between segments of sequential code. The theorems shown in Figures 5.11 and 5.12, with the time parameters reduced to a simple form and the use of let bindings, are quite comprehensible. Aside from the assumptions typical of each instruction proof, there are the two DEC28 assumptions, as well as the assumptions that the parameters to the LD instruction both represent nonnegative integers.

The method of assembling theorems for intervals of the microcode sequence instead of pushing through the entire sequence one instruction at at time, is effective in reducing memory demands from large term sizes in the two most complex instructions: AP and RAP.

```
loop1_state =
.... ⊢ (mpc(t_m + 5) = #000111000) ∧
       (s0(t_m + 5) = s0 t_m) ∧
       (s1(t_m + 5) = s1 t_m) ∧
       (s2(t_m + 5) = s2 t_m) ∧
       (s3(t_m + 5) = s3 t_m) ∧
       (memory(t_m + 5) = memory t_m) ∧
       (s(PRE(t_m + 5)) = s(PRE t_m)) ∧
       (e(PRE(t_m + 5)) = e(PRE t_m)) ∧
       (c(PRE(t_m + 5)) = c(PRE t_m)) ∧
       (d(PRE(t_m + 5)) = d(PRE t_m)) ∧
       (free(PRE(t_m + 5)) = free(PRE t_m)) ∧
       (arg(PRE(t_m + 5)) = DEC28(atom_bits(arg(PRE t_m)))) ∧
       (x1(PRE(t_m + 5)) = cdr_bits(memory t_m(x1(PRE t_m)))) ∧
       (x2(PRE(t_m + 5)) = x2(PRE t_m)) ∧
       (buf1(PRE(t_m + 5)) = DEC28(atom_bits(arg(PRE t_m)))) ∧
       (buf2(PRE(t_m + 5)) = buf2(PRE t_m)) ∧
       (mar_pins(PRE(t_m + 5)) = x1(PRE t_m)) ∧
       (car(PRE(t_m + 5)) = car(PRE t_m)) ∧
       (parent(PRE(t_m + 5)) = parent(PRE t_m)) ∧
       (root(PRE(t_m + 5)) = root(PRE t_m)) ∧
       (y1(PRE(t_m + 5)) = y1(PRE t_m)) ∧
       (y2(PRE(t_m + 5)) = y2(PRE t_m)) ∧
       (rmem_pin(PRE(t_m + 5)) = F) ∧
       (write_bit_pin(PRE(t_m + 5)) = T) ∧
       (flag0_pin(PRE(t_m + 5)) = F) ∧
       (flag1_pin(PRE(t_m + 5)) = F)

loop1_nonmajor =
.... ⊢ !t'_m.
       (PRE t_m) < t'_m ∧ t'_m < (t_m + 5) ==>
       ¬is_major_state mpc t'_m

Assumptions:
[ Clock_constraint SYS_clock
; ^SYS_imp
; ((atom_bits(arg(PRE t_m)) = ZERO28) = F)
; (mpc t_m = #000111000)
]
```

Figure 5.9: Theorems for one iteration through loop1

```
loop1_nth_state =
.... |- !n t_m.
        ¬NEG(iVal(Bits28(atom_bits(arg(PRE t_m)))))
        ==>
        n <= (pos_num_of(iVal(Bits28(atom_bits(arg(PRE t_m))))))
        ==>
        (mpc t_m = #000111000)
        ==>
        (mpc(t_m + (5 * n)) = #000111000) ∧
        (s0(t_m + (5 * n)) = s0 t_m) ∧
        (s1(t_m + (5 * n)) = s1 t_m) ∧
        (s2(t_m + (5 * n)) = s2 t_m) ∧
        (s3(t_m + (5 * n)) = s3 t_m) ∧
        (memory(t_m + (5 * n)) = memory t_m) ∧
        (arg(PRE(t_m + (5 * n))) =
          nth n(DEC28 o atom_bits)(arg(PRE t_m))) ∧
        (buf2(PRE(t_m + (5 * n))) = buf2(PRE t_m)) ∧
        (s'(PRE(t_m + (5 * n))) = s'(PRE t_m)) ∧
        (e'(PRE(t_m + (5 * n))) = e'(PRE t_m)) ∧
        (c'(PRE(t_m + (5 * n))) = c'(PRE t_m)) ∧
        (d'(PRE(t_m + (5 * n))) = d'(PRE t_m)) ∧
        (free(PRE(t_m + (5 * n))) = free(PRE t_m)) ∧
        (x1(PRE(t_m + (5 * n))) =
          nth n(cdr_bits o (memory t_m))(x1(PRE t_m))) ∧
        (x2(PRE(t_m + (5 * n))) = x2(PRE t_m)) ∧
        (car(PRE(t_m + (5 * n))) = car(PRE t_m)) ∧
        (parent(PRE(t_m + (5 * n))) = parent(PRE t_m)) ∧
        (root(PRE(t_m + (5 * n))) = root(PRE t_m)) ∧
        (y1(PRE(t_m + (5 * n))) = y1(PRE t_m)) ∧
        (y2(PRE(t_m + (5 * n))) = y2(PRE t_m))

loop1_nth_nonmajor =
.... |- !n t_m.
        ¬NEG(iVal(Bits28(atom_bits(arg(PRE t_m)))))
        ==>
        n <= (pos_num_of(iVal(Bits28(atom_bits(arg(PRE t_m))))))
        ==>
        (mpc t_m = #000111000)
        ==>
        (!t'. (PRE t_m) < t' ∧ t' < (t_m + (5 * n))
              ==>
              ¬is_major_state mpc t')
```

Figure 5.10: Theorems for n iterations through loop1

```
LD_State =
........⊢ let m =
            pos_num_of
            (iVal
             (Bits28
              (atom_bits(memory tₘ
                (car_bits(memory tₘ
                  (car_bits(memory tₘ
                    (cdr_bits(memory tₘ(c tₘ)))))))))))
        in
        let n =
            pos_num_of
            (iVal
             (Bits28
              (atom_bits(memory tₘ
                (cdr_bits(memory tₘ
                  (car_bits(memory tₘ
                    (cdr_bits(memory tₘ(c tₘ)))))))))))
        in
        let mem1 =
            Store14 (free tₘ)
                    (bus32_cons_append #00 RT_CONS
                    (car_bits
                     (memory tₘ
                      (nth n
                       (cdr_bits o(memory tₘ))
                       (car_bits(memory tₘ(nth m
                         (cdr_bits o(memory tₘ))(e tₘ)))))))
                    (s tₘ))
                    (memory tₘ)
        in
        ((mpc(tₘ + (40 + (5 * (m + n)))) = #000101011) ∧
         (memory(tₘ + (40 + (5 * (m + n)))) = mem1) ∧
         (s(tₘ + (40 + (5 * (m + n)))) = free tₘ) ∧
         (e(tₘ + (40 + (5 * (m + n)))) = e tₘ) ∧
         (c(tₘ + (40 + (5 * (m + n)))) =
         cdr_bits(mem1(cdr_bits(mem1(c tₘ))))) ∧
         (d(tₘ + (40 + (5 * (m + n)))) = d tₘ) ∧
         (free(tₘ + (40 + (5 * (m + n)))) =
         cdr_bits(memory tₘ(free tₘ))))
```

Figure 5.11: Microprogramming stage *State* theorem for LD instruction

```
LD_Next =
........ ⊢ let m =
                pos_num_of
                (iVal
                 (Bits28
                  (atom_bits(memory t_m
                    (car_bits(memory t_m
                      (car_bits(memory t_m
                        (cdr_bits(memory t_m(c t_m)))))))))))
            in
            let n =
                pos_num_of
                (iVal
                 (Bits28
                  (atom_bits(memory t_m
                    (cdr_bits(memory t_m
                      (car_bits(memory t_m
                        (cdr_bits(memory t_m(c t_m)))))))))))
            in
            Next t_m
                (t_m + (40 + (5 * (m + n))))
                (is_major_state mpc)
```

Figure 5.12: Microprogramming stage *Next* theorem for LD instruction

5.6 Liveness

If the SECD system initially reaches a *major state*, and if every time it starts in a *major state* all possible paths eventually return it to a *major state*, then the temporal abstraction function from the coarse granularity of time to the medium granularity is total. This important property is essential for the last stage of the proof.

The function state_abs is well-defined only when one of four values is in the MPC9 register, and these are precisely the four values for which the temporal abstraction predicate is_major_state mpc holds. Unfortunately, one cannot prove that the abstraction predi-

cate holds at the abstracted time, *i.e.*(is_major_state mpc) when
(is_major_state mpc))t_c, or in general (P when P) t (which is
definitionally equivalent to P(TimeOf P t)). This limitation arises
from the use of the SELECT operator @ in defining the temporal
abstraction function TimeOf[7], given in Figure 4.15. This obstacle is
overcome by proving a "liveness" property for the predicate used for
the abstraction (is_major_state mpc). Liveness as defined by Inf
states that the predicate is true infinitely often.

```
⊢ Inf f = !t. ?t'. (t<t') ∧ (f t')
```

The proof of this property is simplified by using the following theorems.

```
Inf_thm =
⊢ !f. (?t'. 0 < t' ∧ f t') ∧
      (!t. f t ==> (?t'. t < t' ∧ f t'))
      ==> Inf f

Next_exists_thm =
⊢ !t t1 f. Next t t1 f ==> (?t'. t < t' ∧ f t')
```

The first limits the proof requirement to showing that a *major state*
is reached initially, and every time the machine starts in a *major
state* it will reach a *major state* again in the future. This reduces the
number of starting points to four, instead of every possible machine
state (consisting of 400+ values that mpc can have). The microprogramming stage theorems defining the Next times the temporal
abstraction function holds are used for each branch of the proof. The
second theorem expresses that part of the definition of Next that satisfies the proof requirement for times starting in *major states*.

---

[7]From the defining axiom for the SELECT operator SELECT_AX:
⊢∀(P:*->bool) (x:*). P x ==> P($@ P), it is necessary to show that P holds
at some value x in order to derive that P($@ P) holds. See [24] for further
description.

```
liveness =
[ clock_constraint SYS_Clocked
; ^SYS_imp
; valid_program_constraint memory mpc button_pin s e c d
; reserved_words_constraint mpc memory
; well_formed_free_list memory mpc free s e c d
; DEC28_assum1
; DEC28_assum2
]
  ⊢ Inf(is_major_state mpc)
```

The proof is achieved by a series of case splits: first using Inf_thm it is split into time 0_c and times t_c such that is_major_state mpc t_c. The latter is split into the four *major states* using the definition of is_major_state, and finally valid_program_constraint divides the possible transitions from top_of_cycle state into 18 cases, each of which is solved by the appropriate *_Next theorem from the previous stage. This was the simplest stage of the proof, requiring under 900 primitive inferences.

5.7 Computations across abstraction

The microprogramming stage provides a set of theorems that define the lower level view of the effect of computation of each instruction, giving new values for each of the s, e, c, d, and free state variables and the low level memory in terms of low level operations on the memory and register contents. The *specification* for each transition defines the new values for the same state variables and the *abstract* memory in terms of abstract memory operations on the previous state values and abstract memory. To complete the verification, it must be proved for each transition that the low level computation corresponds to the transition on the abstracted state specified by the top level specification.

5.7.1 Primitive abstract operations

Relating transitions at the specification and implementation levels requires that the individual operations defining the transition be related, and for this purpose, theorems must be derived relating the abstract memory operations to the lower level operations. A sampling of these theorems follows.

On page 94 we alluded to the need to ensure that the abstraction function from low level memories to abstract memories is total. This requires that the value returned by (Mem_Range_Abs o memory) is within the representative type for abstract memories. By first proving this fact, we can ensure the mapping an abstracted memory back to its representative type is well defined. Recall the definition:

```
mem_abs(memory:word14->word32) =
    ABS_mfsexp_mem(Mem_Range_Abs o memory).
```

The idea is captured in the following theorem.

```
⊢ REP_mfsexp_mem(ABS_mfsexp_mem(Mem_Range_Abs o memory)) =
  Mem_Range_Abs o memory
```

This result is used repeatedly to derive theorems about the relation between abstract operations defined using the REP_mfsexp_mem function on memories abstracted from the lower level.

We begin with the structural operations.

```
car_cdr_mem_abs_lemma =
⊢ is_cons(memory v) ==>
  (!x:word14.
    (M_Car(v,mem_abs memory,x) = car_bits(memory v)) ∧
    (M_Cdr(v,mem_abs memory,x) = cdr_bits(memory v)))

M_Cons_unfold_1_thm =
⊢ reserved_words_constraint mpc memory ∧
  well_formed_free_list memory mpc free s e c d ==>
  (!t_m.
    (state_abs(mpc t_m) = top_of_cycle) ==>
    (!v w:word14.
      let (mem1:word14->word32) =
          (Store14(free t_m)
                  (bus32_cons_append #00 RT_CONS v w)
                  (memory t_m))
      in
      M_Cons(v,w,mem_abs(memory t_m),free t_m) =
      free t_m, mem_abs mem1, cdr_bits(memory t_m(free t_m))))
```

The antecedents in each theorem express conditions under which the low level operations correspond to well-defined operations on abstract memory objects. The constraints of the latter theorem are needed to ensure a cell is available in the free list. The proofs of the theorems are achieved by unfolding definitions of the operations and the mem_abs function, applying the theorems about the totality of the mem_abs function to eliminate the REP and ABS functions, unfolding the definition of Mem_Range_Abs, and using the distinctness of record types.

The complexity generated by successive memory writes makes it desirable to prove a set of theorems that unfold the M_Cons operation when performed on already modified memories. We show only the second of the four such theorems. Theorems M_Cons_unfold_3_thm and M_Cons_unfold_4_thm are similar.

```
M_Cons_unfold_2_thm =
⊢ reserved_words_constraint mpc memory ∧
  well_formed_free_list memory mpc free s e c d ==>
  (!t_m.
    (state_abs(mpc t_m) = top_of_cycle) ==>
    (!(v w:word14) (z:word32).
      let mem1 = (Store14(free t_m) z (memory t_m))
      in
      let mem2 = (Store14 (cdr_bits(memory t_m(free t_m)))
                  (bus32_cons_append #00 RT_CONS v w) mem1)
      in
      M_Cons(v,w,mem_abs mem1, cdr_bits(memory t_m(free t_m))) =
      cdr_bits(memory t_m(free t_m)),
      mem_abs mem2,
      cdr_bits(mem1(cdr_bits(memory t_m(free t_m)))))))
```

Extractor functions are straightforward. The last of the following functions combines two operations that are used to fetch the machine instruction, an often used operation.

```
atom_number_mem_abs_lemma =
⊢ is_number(memory v) ==>
  (!x.
    M_atom_of(v,mem_abs memory,x) =
    Int(iVal(Bits28(atom_bits(memory v)))))

atom_symbol_mem_abs_lemma =
⊢ is_symbol(memory v) ==>
  (!x.
    M_atom_of(v,mem_abs memory,x) =
    Symb(Val(Bits28(atom_bits(memory v)))))

number_mem_abs_lemma =
⊢ is_number(memory v) ==>
  (!x.
    M_int_of(v,mem_abs memory,x) =
    iVal(Bits28(atom_bits(memory v))))

opcode_mem_abs_lemma =
⊢ is_cons(memory v) ==>
  is_number(memory(car_bits(memory v))) ==>
  (!x.
    M_int_of(M_CAR(v,mem_abs memory,x)) =
    iVal(Bits28(atom_bits(memory(car_bits(memory v))))))
```

Arithmetic operations add another twist. The top level operations, such as M_Add, are defined by operations on the finite cyclic group of integers that can be represented by 28 bits using two's complement representation, typified by modulo_28_add. The low level operations performed by the ALU are defined using the SELECT operator @ on the result of applying the group operation to the integer representation of the :word28 arguments. In order to prove the equivalence of the two operations, one must first show for the value returned that some such :word28 value exists and is, moreover, unique.

The proof of existence begins by proving the bounds of the co-domain of the norm28 function (refer to page 80 for definitions).

```
norm28_bounds =
⊢ !M:integer.
    ((neg(INT(2 EXP 27))) below (norm28 M) ∨
     (neg(INT(2 EXP 27)) = norm28 M)) ∧
    (norm28 M) below (INT(2 EXP 27))
```

Following this, the bounds of representation of an n-bit word are proved, and specialised to the case of 28-bit words.

```
iVal_bound =
⊢ !b:(bool)bus.
    let n = Width b
    in
      (((neg(INT(2 EXP (n - 1)))) below (iVal b) ∨
        (neg(INT(2 EXP (n - 1))) = iVal b)) ∧
       (iVal b) below (INT(2 EXP (n - 1))))

iVal_28_bound =
⊢ ((neg(INT(2 EXP 27))) below (iVal(Bits28 w28)) ∨
   (neg(INT(2 EXP 27)) = iVal(Bits28 w28))) ∧
   (iVal(Bits28 w28)) below (INT(2 EXP 27))
```

The existence of an n-bit representation of an integer in a given range is proved; this result specialises to values in the codomain of norm28, and finally specialises to 28-bit words.

```
wordn_int_rep_exists =
⊢ !(M:integer) (m:num).
    ((neg(INT(2 EXP m))) below M ∨ (neg(INT(2 EXP m)) = M)) ∧
    M below (INT(2 EXP m)) ==>
    (?b. (Width b = SUC m) ∧ (iVal b = M))

exists_norm28_rep = ⊢ !i. ?b. (Width b = 28) ∧ (iVal b = norm28 i)

exists_word28_rep = ⊢ !i. ?w28. iVal(Bits28 w28) = norm28 i
```

The uniqueness of the 28-bit value is established first by proving the iVal abstraction function is one-one when applied to equal length bit arguments, and this result is specialised to 28-bit values.

```
iVal_11 =
⊢ !b b'. (Width b' = Width b) ==> ((iVal b = iVal b') = (b = b'))

iVal_Bits28_11 = ⊢ (iVal(Bits28 x) = iVal(Bits28 y)) = (x = y)
```

With these results in hand, the arithmetic operations are solved.

```
M_Add_unfold_lemma =
⊢ is_number(memory x) ∧ is_number(memory y) ∧
  is_cons(memory free) ==>
  (M_Add(x,y,mem_abs memory,free) =
   free,
   mem_abs
   (Store14
    free
    (ADD28((atom_bits o memory)x)((atom_bits o memory)y))
    memory),cdr_bits(memory free))

M_Sub_unfold_lemma =
⊢ is_number(memory x) ∧ is_number(memory y) ∧
  is_cons(memory free) ==>
  (M_Sub(x,y,mem_abs memory,free) =
   free,
   mem_abs
   (Store14
    free
    (SUB28((atom_bits o memory)x)((atom_bits o memory)y))
    memory),cdr_bits(memory free))

M_Dec_unfold_lemma =
⊢ is_number(memory x) ∧ is_cons(memory free) ==>
  (M_Dec(x,mem_abs memory,free) =
   free,mem_abs(Store14 free(DEC28((atom_bits o memory)x))memory),
   cdr_bits(memory free))

M_Leq_unfold_lemma =
⊢ is_number(memory x) ∧ is_number(memory y) ==>
  (M_Leq(x,y,mem_abs memory,free) = LEQ_prim(memory x)(memory y))

M_Eq_unfold_lemma =
⊢ (!x. garbage_bits(memory x) = #00) ==>
  is_atom(memory x) ∨ is_atom(memory y) ==>
  (M_Eq(x,y,mem_abs memory,free) = (memory x = memory y))
```

Lastly, theorems for the atomic predicate operations and a destructive replace operations are given.

```
M_Atom_unfold_lemma =
⊢ M_Atom(v,mem_abs memory,free) = is_atom(memory v)

Rplaca_unfold_lemma =
⊢ !a memory.
   is_cons(memory a) ==>
   (!v free.
    M_Rplaca(a,v,mem_abs memory,free) =
    a,
    mem_abs
    (Store14
    a
    (bus32_cons_append #00 RT_CONS v(cdr_bits(memory a)))
    memory),free)
```

5.7.2 Multiple memory writes and locality

The occurrence of more than one write operation to the memory during the execution of a single machine instruction contributes greatly to the complexity of the last stage of the verification. In the simulated execution of the μ-programming stage of proof, the multiple writes generate enormous terms defining the state of the memory, and values defined in terms of that memory. At the final stage, the problem is more concerned with the locality of effect of the writes, and the retention of data and structure in the rest of the memory. In simple terms, it is necessary to be able to determine that under the constrained circumstances in which the memory is altered, by writing only to otherwise unused cells in the free list, no other existing data structure or object (reachable from any register in the chip) is altered. The idea, although reasonably simple to write in English, proves rather complex to define formally.

It was hoped that a macrotheorem could be defined, which would state that for up to four writes to memory in any sequence, and subject to the constraints on the free list and memory, all existing data structures were the same in the updated memory as they were in the original memory. An attempt to define such a theorem using the `path` function produced theorems of the following form (shown for two writes to memory).

```
⊢ (is_cons(mem free)) ∧
  (is_cons(mem(cdr_bits(mem free)))) ∧
  (mem NIL_addr =
   bus32_symb_append #00000000000000000000000000000) ==>
  !v.
  (nonintersecting mem free v) ==>
   !p cell1 cell2.
    let mem1 = Store14 free cell1 mem
    in
    let mem2 = Store14 (cdr_bits(mem free)) cell2 mem1
    in
    ((path mem2 v p = path mem v p)
     ∧
     (mem2(path mem v p) = mem(path mem v p))
     ∧
     (mem2(path mem2 v p) = mem(path mem v p)))
```

The application of these theorems proves exceedingly tedious however. Deriving any property of any location in a data structure specified by the path function first required establishing that each cell along the path in the updated memory is a cons cell if it was so in the original memory. Instead of repeatedly deriving this result for specific instances, a somewhat large suite of theorems covering the required data structure locations is derived, for each of one through four memory writes. The theorems for two writes are typical.

```
Store14_root_2_lemma =
⊢ is_cons(mem free) ∧ is_cons(mem(cdr_bits(mem free))) ∧
  (mem NIL_addr =
   bus32_symb_append #00000000000000000000000000000) ==>
  (!v. nonintersecting mem free v ==>
       (!cell1 cell2.
        let mem1 = Store14 free cell1 mem
        in
        let mem2 = Store14(cdr_bits(mem free))cell2 mem1
        in
        (mem2 v = mem v)))
```

```
Store14_car_cdr_2_lemma =
⊢ is_cons(mem free) ∧ is_cons(mem(cdr_bits(mem free))) ∧
  (mem NIL_addr =
   bus32_symb_append #00000000000000000000000000000000) ==>
  (!v. is_cons(mem v) ∧ nonintersecting mem free v ==>
       (!cell1 cell2.
         let mem1 = Store14 free cell1 mem
         in
         let mem2 = Store14(cdr_bits(mem free))cell2 mem1
         in
         ((mem2(cdr_bits(mem2 v)) = mem(cdr_bits(mem v))) ∧
          (mem2(car_bits(mem2 v)) = mem(car_bits(mem v))))))

Store14_caar_cdar_2_lemma =
⊢ is_cons(mem free) ∧ is_cons(mem(cdr_bits(mem free))) ∧
  (mem NIL_addr =
   bus32_symb_append #00000000000000000000000000000000) ==>
  (!v. is_cons(mem v) ∧ is_cons(mem(car_bits(mem v))) ∧
       nonintersecting mem free v ==>
       (!cell1 cell2.
         let mem1 = Store14 free cell1 mem
         in
         let mem2 = Store14(cdr_bits(mem free))cell2 mem1
         in
         ((mem2(cdr_bits(mem2(car_bits(mem2 v)))) =
           mem(cdr_bits(mem(car_bits(mem v))))) ∧
          (mem2(car_bits(mem2(car_bits(mem2 v)))) =
           mem(car_bits(mem(car_bits(mem v))))))))

Store14_cadr_cddr_2_lemma =
⊢ is_cons(mem free) ∧ is_cons(mem(cdr_bits(mem free))) ∧
  (mem NIL_addr =
   bus32_symb_append #00000000000000000000000000000000) ==>
  (!v. is_cons(mem v) ∧ is_cons(mem(cdr_bits(mem v))) ∧
       nonintersecting mem free v ==>
       (!cell1 cell2.
         let mem1 = Store14 free cell1 mem
         in
         let mem2 = Store14(cdr_bits(mem free))cell2 mem1
         in
         ((mem2(cdr_bits(mem2(cdr_bits(mem2 v)))) =
           mem(cdr_bits(mem(cdr_bits(mem v))))) ∧
          (mem2(car_bits(mem2(cdr_bits(mem2 v)))) =
           mem(car_bits(mem(cdr_bits(mem v))))))))
```

These theorems can be resolved with constraint-derived assumptions and used to establish the constancy of data structure components.

5.7.3 On with the proof

The top goal for the final proof (Figure 5.2) splits into two parts: the state when the machine first reaches a *major state*, and the state of the machine when it is next in a *major state*, in terms of its state in the previous *major state* (*i.e.* at time t_c). The first goal is quite simply solved by the theorem for the initial state given on page 122, stating that the system is initially in the *idle* state.

The second part of the proof is partitioned into twenty-one subparts: one for each of the eighteen (implemented) machine instructions originating in the *top_of_cycle* state, and one for each of the *idle, error1* and *error2* states. A theorem is proved for each part, relating the abstract top level specification transition to the RT level transition effected by the machine, under the assumption of the appropriate starting state and the constraints that cause the particular transition to be selected, and also the applicable constraints of the top level correctness goal.

The second part may be split into a set of subgoals, corresponding to each transition defined in the SYS_spec. The transitions are determined by the state at time t_c, which is a function of the mpc at TimeOf(is_major_state mpc)t_c. Using the liveness theorem from the previous stage, with the theorem TimeOf_TRUE:

⊢!f. Inf f ==> (!n. f (TimeOf f n)),

we obtain the result that is_major_state mpc is true at all points of the coarser granularity of time.

⊢ !t_c. is_major_state mpc (TimeOf (is_major_state mpc) t_c)

With this result, the state abstraction function is well defined, and we know that mpc has one of four values at all points in the coarser granularity of time.

Theorems for the correctness of each top level transition are typified by the LDF transition theorem in Figure 5.13. The first five

```
correctness_LDF =
[ clock_constraint SYS_Clocked
; ~SYS_imp
; reserved_words_constraint mpc memory
; well_formed_free_list memory mpc free s e c d
; mpc(TimeOf(is_major_state mpc)t_c) = #000101011
; Inf(is_major_state mpc)
; is_cons(memory(TimeOf(is_major_state mpc)t_c)
          (c(TimeOf(is_major_state mpc)t_c)))
; is_number(memory(TimeOf(is_major_state mpc)t_c)
            (car_bits(memory(TimeOf(is_major_state mpc)t_c)
                     (c(TimeOf(is_major_state mpc)t_c)))))
; atom_bits(memory(TimeOf(is_major_state mpc)t_c)
            (car_bits(memory(TimeOf(is_major_state mpc)t_c)
                     (c(TimeOf(is_major_state mpc)t_c))))) =
  #00000000000000000000000000000011
; is_cons(memory(TimeOf(is_major_state mpc)t_c)
          (cdr_bits(memory(TimeOf(is_major_state mpc)t_c)
                   (c(TimeOf(is_major_state mpc)t_c)))))
]
 ⊢ ((s                   when (is_major_state mpc))(SUC t_c),
    (e                   when (is_major_state mpc))(SUC t_c),
    (c                   when (is_major_state mpc))(SUC t_c),
    (d                   when (is_major_state mpc))(SUC t_c),
    (free                when (is_major_state mpc))(SUC t_c),
    ((mem_abs o memory)  when (is_major_state mpc))(SUC t_c),
    ((state_abs o mpc)   when (is_major_state mpc))(SUC t_c) =
    LDF_trans ((s               when (is_major_state mpc)) t_c,
              (e                when (is_major_state mpc)) t_c,
              (c                when (is_major_state mpc)) t_c,
              (d                when (is_major_state mpc)) t_c,
              (free             when (is_major_state mpc)) t_c,
              ((mem_abs o memory) when (is_major_state mpc)) t_c)
```

Figure 5.13: The Correctness result for the LDF instruction

hypotheses correspond to those of the LDF_state theorem. The liveness result is assumed in the Inf(...) hypothesis, since only part of the valid_program_constraint is present, and thus liveness cannot be derived for the single case. The remaining hypotheses match the applicable portion of valid_program_constraint for the LDF instruction branch. The first three of these correspond to the single last hypothesis of the LDF_state theorem "opcode_bits(memory t_m(car_bits(memory t_m(c t_m)))) = #000000011", and illustrate the different degree of detail demanded by the memory abstraction. It is not enough to know that the 9 lower bits of the next instruction constitutes the LDF opcode. Accessing this cell in the abstracted memory requires that the first cell in the C stack is a cons cell, and the cell containing the opcode is a number cell, and finally that all 28 bits are known, as the opcode is defined by the Val abstraction of all 28 bits. The final hypothesis is similarly part of the data structure requirement for the abstract memory operations to be defined. These detailed constraints effect the left hand side pattern matching inherent in the simplest state transition definition of the machine.

The proof of the correctness_LDF theorem proceeds in the following steps (and likewise for the other instruction theorems):

1. Derive that the machine is in *top_of_cycle* state at time t_c, using the fifth assumption (value of the mpc).

```
[ "state_abs(mpc(TimeOf(is_major_state mpc)t_c)) =
   top_of_cycle" ]
```

2. Derive the value of the lowest 9 bits of the 28 bit instruction code from the second last assumption, for resolving with the less specific assumption of LDF_state and LDF_Next.

```
[ "opcode_bits
    (memory
    (TimeOf(is_major_state mpc)t_c)
    (car_bits
     (memory
      (TimeOf(is_major_state mpc)t_c)
      (c(TimeOf(is_major_state mpc)t_c))))) =  #000000011" ]
```

3. Using the definition of **when**, and theorem **LDF_Next**, transform expressions of the form (s when (is_major_state mpc))(SUC t_c) to s((TimeOf (is_major_state mpc)t_c) + 26).

4. For efficiency and readability, introduce the new variable t_c' to abbreviate the term (TimeOf(is_major_state mpc)t_c). The goal, omitting assumptions, looks as follows.

```
s(t_c' + 26),e(t_c' + 26),c(t_c' + 26),d(t_c' + 26),free(t_c' + 26),
mem_abs(memory(t_c' + 26)),state_abs(mpc(t_c' + 26)) =
LDF_trans(s t_c',e t_c',c t_c',d t_c',free t_c',mem_abs(memory t_c'))
```

5. Generate individual constraints on free list cells by resolving the **well_formed_free_list** assumption with the state assumption and **reserved_words_constraint**.

```
[ "is_cons(memory t_c'(free t_c'))" ]
[ "is_cons(memory t_c'(cdr_bits(memory t_c'(free t_c'))))" ]
...
[ "~(free t_c' = cdr_bits(memory t_c'(free t_c')))" ]
[ "~(free t_c' =
    cdr_bits(memory t_c'(cdr_bits(memory t_c'(free t_c')))))" ]
...
[ "memory t_c' NIL_addr =
    bus32_symb_append #00000000000000000000000000000000" ]
```

6. Obtain assumption for the nonintersection of the free list and the data structure pointed to by the c register contents, using the properties of the **well_formed_free_list**.

```
[ "nonintersecting(memory t_c')(free t_c')(c t_c')" ]
```

7. Rewrite with the definition of **LDF_trans**. The goal now looks as follows.

```
s(t'_c + 26),e(t'_c + 26),c(t'_c + 26),d(t'_c + 26),free(t'_c + 26),
mem_abs(memory(t'_c + 26))),state_abs(mpc(t'_c + 26)) =
(let cell_mem_free =
        M_Cons_tr
        (s t'_c,
         M_Cons_tr(e t'_c,M_CAR(M_CDR(c t'_c,
                                       mem_abs(memory t'_c),
                                       free t'_c))))
 in
    cell_of cell_mem_free,e t'_c,
    M_Cdr(M_CDR(c t'_c,mem_free_of cell_mem_free)),d t'_c,
    free_of cell_mem_free,mem_of cell_mem_free,top_of_cycle)
```

8. Repeatedly for each *let* binding do the following:

   - Eliminate all the M_Car, M_Cdr, M_CAR, and M_CDR operations using the theorem car_cdr_mem_abs_lemma.

   - Rewrite M_Cons_tr if present.

   - Eliminate the M_Cons operation by rewriting with the appropriate M_Cons_unfold_*_lemma, where * is the occurrence number, between 1 and 4, of this M_Cons operation in the current sequence.

   - Rewrite the *let* binding.

   - Rewrite any of the selector functions cell_of, mem_of, free_of and mem_free_of that have a triple argument.

Continue until all *let* bindings and abstract memory operations are eliminated. The right hand side of the goal equation, which represents the specification for the instruction, is now expressed entirely in low level memory operations.

9. Rewrite with LDF_state, expanding the left hand side of the goal equation to the value resulting from the machine execution.

10. Rewrite the state_abs function applied to the final mpc content to get a named state.

11. Apply REFL_TAC (*i.e.* the terms are now identical).

The pattern of the proof is consistent for all theorems except those having a conditional branch (including the SEL, ATOM, EQ, LEQ, *idle*, *error0*, and *error1* transitions). For these proofs, a split is made on the condition before the 3rd step can be accomplished.

The use of many of the theorems, particularly those relating operations at different levels, and the locality of memory effects, requires resolution with several of the goal assumptions. The implicative form of the theorems, where the antecedent is a conjunction of several terms, can give rise to literally hundreds of resolvents. An effective means to improve the efficiency of this step is to first put the theorems into a canonical form, using the HOL forward inference rule *IMP_CANON*. Given a theorem in such a form, matching each antecedent will be attempted in sequence, instead of first generating several canonical forms with different orderings of the nested antecedents.

One useful note on the derivation of the proof concerns the complexity and size of intermediate goal terms generated. Once again the size alone becomes an obstacle, as the state of the goal is difficult for a frail human to comprehend. Just as *let* bindings have been used to simplify the form of theorems in this work, a comparable device is employed in interactive proofs, demonstrated in the 4th step of the proof. A new variable is introduced to represent a complex or frequently occurring subterm of the goal or assumptions, and substitute the variable for all occurrences, including assumptions, while adding an assumption giving the value of the new variable as the subterm. The tactic is called ABBREV_TAC, and has been included in the "contrib" library distributed with the HOL system. The original state of the goal is recoverable by discarding the assumption after substituting the value for all occurrences of the introduced variable. Even when the abbreviation needs to be undone in order to effect a tactic successfully, it is a useful means of reducing an impossibly complex goal to a simpler form in which patterns can be detected, and strategies devised. Additionally, it is a quick and simple substitution, which is highly desirable when the term size makes most tactics exceedingly slow.

5.7.4   Final correctness result and proof

```
SECD_implements_specification =
[ clock_constraint SYS_Clocked
; SYS_imp
; well_formed_free_list memory mpc free s e c d
; reserved_words_constraint mpc memory
; valid_program_constraint memory mpc button_pin s e c d
; Garbage_bits_constraint memory mpc
]
⊢ SYS_spec
     ((mem_abs o memory) when (is_major_state mpc))
     (s when (is_major_state mpc))
     (e when (is_major_state mpc))
     (c when (is_major_state mpc))
     (d when (is_major_state mpc))
     (free when (is_major_state mpc))
     (button_pin when (is_major_state mpc))
     ((state_abs o mpc) when (is_major_state mpc))
```

Figure 5.14: Top level correctness theorem

The proof of the final theorem shown in Figure 5.14, is understandably straightforward following the completion of proofs of correctness for each possible transition. The proof steps are as follows.

1. Rewrite with the definition of SYS_spec.

2. Split into initial case (trivially solved) and other case.

3. Resolve the liveness theorem and add as an assumption.

4. Uncurry and perform BETA conversion on the specification.

5. Split into cases for each of the 4 major states.

6. 3 of the 4 cases are solved in a straightforward manner using the previously described theorems.

7. The *top_of_cycle* state must be split into the 18 subcases. Begin by rewriting with the definition of Next. The case split is performed using the disjunctive valid_program_constraint.

8. The `valid_program_constraint` is given as an argument to a proof function, which repeatedly splits off one disjunct of the constraint, and solves the goal under that as an assumption, using one of the preproved theorems. The rest of the task of the proof function consists of reducing arithmetic equalities to true or false. These arise from the large case statement of the **NEXT** function, which repeatedly compares the machine instruction (an integer) with integer constants. In order to do this methodically, a theorem that the < relation holds for successive number pairs (up to 20<21) is proved. The **iVal** abstraction of each operand value is also derived. These theorems are used to reduce each conditional operator in the goal to a boolean constant, using the transitivity of **LESS** (the integer version of <), and deriving inequality using the built-in theorem **LESS_NOT_EQUAL**.

9. At the completion of the tactical proof, the two DEC28 assumptions, which are still in the set of assumptions, are eliminated by supplying the matching proved theorems.

5.8 Summary

This chapter has presented an overview of the proof of the correctness theorem relating the RT and top level specifications of the SECD system. Constraints limiting the scope restrict the chip to normal mode of operation and a properly configured memory. This latter constraint involves the values in reserved memory locations, the form of the free list, the value of the garbage bits of each record, and the permissible machine codes and system state associated with each. The last item effectively defines the pattern matching inherent in the informal state transitions defining the abstract machine in Table 2.3. The constraints on the free list concern both its structure, and its separation from data structures in the memory. By requiring a minimum number of records in the free list, consideration of garbage collector correctness is deferred.

The refinement of the constraints evolved as the proof proceeded. What in retrospect appears as an obvious collection of conditions is the result of incremental extensions, reflecting a strong desire to minimise the constraints to those necessary to achieve the proof. The final versions capture complicated conditions quite elegantly, and express some of the properties central to a specification for the garbage collector.

The proof presented in this chapter compares in size with the largest of previous efforts. The problem size relates largely to the complexity of individual instruction semantics, as well as the system size.

The staged approach to the proof owes much to previous efforts in microprocessor verification, particularly those of Gordon [20], Cohn [12, 13], and Joyce [34, 35], as well as the work of Melham [43] and Dhingra [16] on temporal abstraction. The methodology at each stage differs mainly from prior work in the sheer size of proof effort required, demanding much less user intervention at several stages, and the generality of the approach.

Problem size dominates the proof strategy at each stage. The first stage of proof has enormous terms, which are many times larger than can be displayed on a workstation screen. Term size also limits the use of powerful tactics such as rewriting, since intermediate steps in the tactic can exhaust available memory, and cause failures. Limiting the tools to the most primitive rules and tactics makes otherwise trivial proofs both tedious and time consuming.

The phase proof stage has to cope with both large size terms and large numbers of theorems. The high overhead required to prove inequality of $word_n$ constants leads to exhaustive case theorems for the possible values, which are then used repeatedly as required by other proofs. User intervention is minimised by use of a proof function, although many hours of work went into the latter's design. Any alteration to the proof function required regenerating all theorems, impeding progress for a day and a half. The theorems have to be divided among several HOL theories, otherwise once again available memory is exhausted. This has the advantage of more efficient theorem retrieval at later stages.

The microprogramming stage offers some of the more interesting proof challenges. The number of theorems generated is still large enough to demand minimising user intervention, but the stronger motivation is the length of time required for each proof, much of which is spent retrieving phase theorems. Timing information about the execution time for each sequence is part of the proof result, rather than information supplied by the user.

Term size becomes a factor as the number of records written to the memory during a single machine instruction transition rises. The expression size appears to grow exponentially in this number, so that some sequences cannot be proved with the original proof function. While the final theorem can be simplified by the use of let expressions, the intermediate terms generated during the proof are so large and complex as to be completely incomprehensible, and only the machine based proof assistant can reliably cope.

The proof of termination of the microcoded loops has not appeared in previous microprocessor proofs in the form shown here[8], although such proofs have long been part of the domain of software verification (for example Gordon [23]).

The liveness stage of proof is made trivial by the design of the constraints and the careful design of the theorems deriving from the microprogramming stage. Only the simple initial transition caused some difficulty, forcing the development of techniques for dealing with the @ operator used in the definition of the temporal abstraction function.

The use of different data types to define state transitions at each level contributes an interesting study in the use of data abstraction. While the benefits in this particular abstract memory data type may be argued, the concept of higher level data structures with a limited number of primitive operations, representing much more complex lower level manipulations, can clearly be useful in presenting information more clearly and concisely. The added proof effort is minimal, since proof of the correspondence between operations must be done only once for each operation, and these results may be used repeatedly.

---

[8] Joyce [35] uses a form of temporal logic in a microprocessor proof with an asynchronous memory interface, utilizing a microcode busy wait loop.

Exploring further

All of the suggested areas for further specification at the end of the previous chapter will pose interesting challenges if carried forward to proofs of correctness. Many will require adjusting the constraints defined in this chapter. A reformulation of the top level correctness goal is suggested in the concluding chapter as well.

Chapter 6

Denouement

The work described in this book has covered a wide spectrum from abstract architecture through VLSI design and layout, to formal specification and formal proof of correctness. We have:

- described the abstract SECD architecture and shown how it supports execution of a high level functional language,

- described the evolution of a realized system specification and hardware implementation,

- presented the formal definitions of the specification and an implementation view of the realized system,

- formalised the constraints under which the system is designed to function, and

- shown the method of achieving a machine proof of correctness that the lower level correctly implements the top level specification.

The end result of the work described is the production of a *partially* verified hardware implementation of a functional architecture. Although many gaps remain between the hardware device and the formal description, some advances in linking formal methods with hardware design have been achieved. The path taken, beginning with the physical design followed by formal representation and partial verification, was a direct result of availability of talents at the start of the project and the long and sometimes painful process of becoming adept at the use of the HOL system. The intimate knowledge of design issues was necessary in the formalisation of representations, but the physical layout did not benefit from the use of formal methods in the design evolution. In several aspects of the design, this lack was apparent at later stages, particularly the odd *write-through* of the ARG register when fetching a machine instruction, and also the storing of a pointer to the computation result in executing the STOP instruction. The former most likely arose from incremental modifications, where originally the instruction was fetched but not stored in a datapath register. The implications of such incremental changes are not always obvious, whereas creating the formal specification drew attention to this feature.

In all fairness, the points made above are minor compared to other problems which were discovered in the physical layout. The fabricated chips have wiring errors in the *shift registers*, and a wiring error in an *XOR* gate in the datapath cell library. Rather than arguing against the use of formal methods, these failures draw attention to the limitations of their use. In representing a physical device, we are proposing that a formal description captures the abstract essence of the device. We need to ensure that the formal description accurately represents the circuit, which in both cases it did not. The use of tools to automatically generate layout, such as described by Slind [52] that transforms HOL specifications into *LSI Logic* gate array net-lists, or tools to check consistency of layout net-lists with their formal representation, could have avoided the problems we encountered. Indeed, even the *Electric* layout tool has a *Mossim* circuit extractor which could have been used to compare with the *Mossim* simulation model. An argument for verification of such tools themselves parallels the argument for the use of formal methods in specifying the design.

Aside from the scope of the work including actual chip design, some innovations have added to previous work on microprocessor verification. The use of an abstract data type and operations thereon to capture the top level behaviour and the resulting proof of the correspondence of the abstract and more primitive operations has provided a more understandable specification at the top level. It is essential that this level be understandable if we are to realistically relate it to the *intention* of the designer.

The top level specification also explicitly includes an initial state requirement, which necessitates lower level constraints on clocking and the *reset* input. Such explicit treatment of constraints is a step forward in providing meaningful information to designers using a product. In hand with the initial state specification, the next state transition specification has been defined more generally, with the explicit time parameter generalised internally in the definition, unlike the single step specification used by Cohn [12, 13]. This produces a clearer correctness statement than the examples cited, particularly so because the temporal relation between the granularity of time in the two levels is expressed entirely in terms of the **when** function applied to the predicate identifying points of time in the finer grain that correspond to points at the coarser grain. This avoids the need for an explicit function to give the time between synchronization points relating the two time granularities. These values in fact are never supplied, but are generated by the proof process, which is sensible when the system complexity is considered.

The proof of the effect of computation of the microcoded *while* loops presents techniques which may be useful in similar control sequences. The approach shows how the principles of proof of while loop constructs in Floyd-Hoare logic are adaptable to hardware, when the explicit time information must be incorporated.

One other difference has been the use of a well-defined multi-bit word data type. This data type has advantages in representing constants and simplicity of defining operations such as subfield extraction. Generic low level components can be defined with parameters of type : (bool)bus, and later instantiated to a particular word size. Proofs of correctness of the generic definition include explicit constraints of equality of word widths of its parameters in many cases. These are easily satisfied when the supplied parameters are of the same $word_n$ type, appropriately converted by a $Bits_n$ function to the : (bool)bus type.

6.1 Putting the Proof Result into Context

The completion of the single level of correctness proof of such a complex device is a substantial accomplishment. Despite the constraints and limitations of the SECD proof of correctness, it still stands as one of the largest completed verifications. Some measure of the size of the task can be obtained by recording the number of primitive inference steps required to define, axiomatize, and verify the two levels of description. Primitive inferences even within the HOL system are not, however, a very reliable indicator as Table 6.1 shows. The development of the HOL system between versions has made significant improvements to the efficiency in several areas, particularly in rewriting which dominates a lot of the proof. The proof was little changed between the versions, merely making those changes that were needed for compatibility with improvements to the system. It should be noted that both versions of the proof were completed using HOL88 built on Allegro Common Lisp.

proof stage	primitive inferences	
	SECD v1.11	SECD v2.0
data type definitions	99,469	50,627
definition	257,048	90,459
unfolding definitions	446,585	74,443
phase proofs	4,901,115	1,008,857
microprogramming proofs	1,145,063	430,921
liveness proof	933	874
final stage	1,304,044	667,760
total	8,154,257	2,319,541

Table 6.1: Comparison of SECD proofs on two versions of HOL

The most substantial previous example is Cohn's proof of the Viper microprocessor [13]. Undertaken on a still earlier version of HOL, and without the final stage of the proof being completed, the Viper proof totaled approximately 6,253,000 primitive inferences. The Viper is a fairly simple 32-bit microprocessor designed for safety critical applications, and is commercially available. It is hard-wired rather than microcoded, and consists of approximately 5000 gates. The Viper proof includes proofs of correctness of the implementation of arithmetic operations, which the SECD proof does not.

The proof of correctness does not purport to offer any complete assurance of correctness of the SECD chip design. The limitations clearly stated in the correctness statement as assumptions exclude, among other things, the behaviour of the on-board garbage collector and test mode of chip operation. Other constraints including the `reserved_words_constraint` and those concerning the value of garbage bits in the memory are straightforward configuration constraints. On the other hand, the `valid_program_constraint` is exceedingly complex, and may or may not be useful in relating the execution of instructions on a machine to the high level programs compiled to machine instructions. Additionally, the complexity of the constraint makes its meaning difficult to comprehend both comprehensively and in detail.

The lowest level of representation is an abstraction of a physical system, and one quite far removed from the physical. Detailed temporal abstractions from a lower level physical model were developed, and these must also be understood if one wants to comprehend the correctness statement. Errors at the layout level in manufactured versions of the chip emphasise that inconsistency with the model is possible.

A very substantial limitation on the assurance gained is the limitation of the correctness statement to single transitions. Extending the verification to sequences of transitions means that the constraints must be shown to hold after each transition as well. Clearly this is not possible for the `free_list_constraint`, unless one wishes to assume an infinite supply of free cells (despite the finite address space!). However, the constraint on the garbage bits should be provably retained, as should the `reserved_words_constraint`. The `valid_program_constraint` would need to be recast, building perhaps instead on some domain of machine code programs generated by a compiler.

The HOL system has proved a reliable platform for such work, although forcing very careful management of proofs at times. One observation has arisen repeatedly through the work: the difficult problem is not the proof of specific theorems, but rather the creation of useful and correct specifications and constraints, and the formulation of a strategy to achieve a result which entails a series of theorems. This highlights the need for specification tools such as Camilleri's tools for executable specification [11]. Becoming sufficiently competent in the use of the HOL system to undertake significant proofs takes considerable time and dedicated effort, but in many cases the management of the proof complexity is far more challenging than the management of the thread of the proof, which is often quite mechanical. Where difficulties arise in completing a proof, it is more often a flaw in the definitions or the constraints, and thus the proof process feeds back to the specification and constraints constantly.

The scope of this project has been extended by other works. An analysis and informal proof of correctness of the compilation algorithm by Simpson [51] is a significant step towards the verified system ideal: verified software, compiled by a verified compiler, being executed on a verified hardware system. One hopes some day that such a system may be used to implement a proof environment as well. More important perhaps than the continuity of trustworthiness at each level gained by verification would be the formal specification interfacing each, so that the abstract domain of algorithms can be related to the finite world of hardware in an explicit manner.

Although no attempt has been made to verify that garbage collection is correctly implemented, some provision is included in the model for this later extension, despite the failure to formalise the meaning of what the garbage collector does. A suitable data type to represent this function may perhaps be found in set theory, where a *state* set may consist of all cells reachable from the state registers, disjoint from the set of cells reachable from the free register. A *cons* operation on the abstract memory moves a cell from the *free* set to the *state* set. The garbage collect function causes the universe of cells to be equal to the disjoint union of the two sets, while not altering the state set. It may be possible to define such sets using the path function, but this has not been explored.

The constraints also limit the verification to normal mode of operation, omitting the test mode use of the shift registers. Once again, the low level model and clocking constraints are designed for this possible extension. In other constraints, such as the initialising of the MPC9 register by constraining the clocks and reset input, we are trading off completeness against utility and simplicity of the specification. We really are unconcerned with how the device may operate under circumstances when these constraints are not in force. The action of the chip when illegal instruction codes are encountered is perhaps an exception. A predictable and traceable recovery may be desirable, but this omission is at least clearly discernible from the valid_program_constraint in the correctness result.

The limitation of the verification to the two upper levels of description of the system has resulted from a limitation of time rather than any difficulty in the problem itself. The extension of the proof to the lower level is expected to be considerably more simple in many respects, since it will be feasible to treat components such as registers and the ALU individually. The datapath composition of these components matches the *RTL* level view almost entirely, hence the composition itself will require little consideration. Given a library of datapath components, the *RT* level is an appropriate level at which to stop in VLSI specification. Similarly, a verification that the low level definition of the *control unit* ROM correctly implements the *RT* ROM should derive from a library ROM model, tailored with a particular transistor layout. The remainder of the *control unit* hierarchy differs between the two levels, and will require a more complex proof effort, but not one which raises new problems.

6.2 Retrospective Improvements

Looking back, there is much that could be improved, particularly in the design of the chip. The lack of concern with speed of operation is explicit, but some aspects of the design can be improved dramatically without increasing complexity. Among the most inefficient features is the instruction fetch operation. With an external RAM, the memory interface is the determining timing constraint, and the timing is further delayed by passing it through the ARG register, which slows the achievable clock rate. The addition of an extra cycle for this operation could well be made up by a higher clock rate.

The imbalance of the clock phases, with only a single logic element separating the output of the latch clocked on ϕ_A from the input of the latch clocked on ϕ_B, versus the entire rest of the chip and memory logic for the other phase, means much more computation occurs during one half of the cycle than the other half. Separating the pair of latches holding the controller state, placing the ϕ_A triggered latch after the ROM for example, could even out this imbalance considerably, and perhaps contribute to a better overall clocking scheme. The datapath registers could be clocked on ϕ_B, thus having both *datapath* and *control unit* changing on the same clock phase.

A modification driven by the formal specification would separate the functionality of the decrement operations of numbers and addresses. The added complexity from duplicating functional components is mitigated by avoiding the devices needed to pad address values with zero's.

Some of the most time consuming parts of the verification are directly attributable to the lack of low level supporting theories, such as the abstraction from a fixed word size data type to integers. This sort of theory required both definition and axiomatization, and is clearly something that should be a built-in feature of any serious hardware specification and verification system. Specifically, standardized abstractions from finite words to types :num and :integer and the associated axiomatizations are required. Fast, and perhaps even insecure, tools that do trivial arithmetic can save considerable time, and yet be replaced by the secure versions for the final proof. Proof development time is too often wasted waiting for arithmetic proofs to complete.

The use of a relational specifications for the STOP instruction points instead of an equational one points to the fact that many of the instruction transitions are over specified, for example by arbitrarily determining the order of non-interfering writes to memory. It should be the intent of the specifier to capture the essential properties of the machine, and not prohibit otherwise acceptable implementations. Combined with the earlier comments on retaining the properties imposed by the constraints, one can envision the form of specification becoming something like the following, wherein the major_state_constraints assure some property of the state under which the machine behaves as specified, and further, this property is retained by the behaviour of the machine.

```
most_general_constraints
⊃
    implementation (state) (inputs) (outputs)
    ⊃
        (startup_constraints
        ⊃
            ((initial_state_spec ((abs o inputs)0,
                                   (abs o state)0,
                                   (abs o outputs)0)
            ∧
            major_state_constraints ((abs o state)0))
        ∧
        (!t. (major_state_constraints ((abs o state)t))
        ⊃
            ((specification_relation ((abs o inputs)t,
                                       (abs o state)t,
                                       (abs o state)(t+1),
                                       (abs o outputs)(t+1))
            ∧
            (major_state_constraints ((abs o state)(t+1))))))
```

Figure 6.1: Skeleton Form of Correctness Statement

6.3 Hardware Verification

So where does this leave formal methods and hardware verification as
a contributor to the development of complex systems? First, we have
shown that a formal specification can not only express the behaviour
of a moderately complex system, but indeed can help clarify what the
behaviour is or should be. Second, using formal inference methods,
we can relate two levels of description of the system, and gain a high
degree of assurance that the lower level model correctly implements
the behaviour of the high level model. By extending this process to a
low enough level, say the model extracted from the circuit layout, or
by automated transformation of formal definitions to layout, we may

decrease the likelihood of wiring errors in the layout. By extending the process to the compiler and software levels, we may be able to better relate algorithms and their execution on hardware.

There remains a vital point. Verification cannot in any way ensure that any hardware device is "correct". At best we deal only with abstract models of things, rather than the actual devices themselves. Even if our model is "correct" in terms of bearing a one to one correspondence to the circuit in the device, the abstraction loses much relevant information. We are constrained immediately by how accurately our formal model captures the behaviour of the device. Errors that occur in the fabrication of the device destroy any accurate relationship in any event. At the other end of the range, we have formal models of some abstract ideas in the head of the designer. The relation between the two is not something that we can ever be sure is accurate. Confidence in the correctness of a specification can be gained by exercising it with a high level (perhaps symbolic) simulator and by subjecting it to public scrutiny (hence the requirement for readability and succinctness).

Of course, these same limitations apply equally to the use of simulation to "verify" design correctness. Simulation at the switch level and above uses the same underlying model as formal specification. The difference lies in *how* the model is used. Simulations run many individual tests to cover all input possibilities. Formal proofs take the same model and the same specifications as are used in the simulation model and manipulate the latter formally to prove the correctness of the design elaboration. Full coverage, that is correctness over all input values, comes automatically. Further, most VLSI designs are regular or contain regular subsystems (e.g. the n-bit adder is a row of 1-bit full adders). Regular sub-systems can be verified using *induction*. Inductive proofs split into two cases; the base case and the inductive case (show the correctness of a sub-system of size n+1 assuming the correctness of a sub-system of size n). So proofs of regular systems do not balloon in length with n, whereas the number of simulation runs required does. Whilst it must be admitted that carrying out a proof is much harder than writing a simulation program, we should also remember that the simulation program will be unproved.

Adopting verification techniques does not impose a new design methodology. Information already present is used in a much more formal way to guarantee the correctness of a design elaboration. Thus verification techniques should embed well into CAD systems.

The methodology, tools, and experience are not yet there, the subject is still in its infancy. There is a strong need for libraries of specifications to be established, and for large case studies to be published so work can proceed towards establishing a robust and reliable technology and automating (part of) it. Despite these current drawbacks, it remains an approach with a promising future. Formal verification should be considered as another weapon in the armoury of hardware designers, particularly useful for showing the correctness of regular systems and for conducting proofs of functionality at the sub-system level and above.

It is hoped that the SECD project has made some contribution, and that its general specification and verification methodology will provide some useful guidelines to others undertaking large examples. It is one of far too few nontrivial completed examples which are widely accessible.

Bibliography

[1] François Anceau. *The Architecture of Microprocessors.* Addison-Wesley Publishing Company, 1986.

[2] W. R. Bevier. A Verified Operating Systems Kernel. Technical Report CLI-11, Computational Logic Inc, Austin, Texas, 1987.

[3] G. Birtwistle, B. Graham, and S-K. Chin. *Hardware Verification in HOL.* in preparation.

[4] G. Birtwistle, J. Joyce, B. Liblong, T. Melham, and R. Schediwy. Specification and VLSI Design. In G. J. Milne and P. A. Subrahmanyam, editors, *Formal Aspects of VLSI Design*, pages 83–97, Amsterdam, 1986. North Holland.

[5] R. S. Boyer and J. S. Moore. *A Computational Logic.* Academic Press, New York, 1979.

[6] R. Bryant. An Algorithm for MOS Logic Simulation. *LAMBDA*, 1980.

[7] W. Burge. *Recursive Programming Techniques.* Addison-Wesley, New York, 1975.

[8] Cambridge Research Center, SRI International, Cambridge, England. *The HOL System: Description*, 1989.

[9] Cambridge Research Center, SRI International, Cambridge, England. *The HOL System: Reference Manual*, 1989.

[10] Cambridge Research Center, SRI International, Cambridge, England. *The HOL System: Tutorial*, 1989.

[11] A.J. Camilleri. *Executing Behavioural Definitions in Higher Order Logic.* PhD thesis, University of Cambridge Computer Laboratory, 1988.

[12] A. J. Cohn. A Proof of Correctness of the VIPER Microprocessor: The First Level. In G. Birtwistle and P. A. Subrahmanyam, editors, *VLSI Specification, Verification and Synthesis*, pages 27–71, Norwell, Massachusetts, 1988. Kluwer. Also University of Cambridge, Computer Laboratory, Tech. Report No. 104.

[13] A. J. Cohn. A Proof of Correctness of the VIPER Microprocessors: The Second Level. In G. Birtwistle and P. A. Subrahmanyam, editors, *Trends in Hardware Verification and Automated Theorem Proving*, pages 1–91, New York, 1989. Springer Verlag.

[14] Dan Craigen. Position Paper for FM89. Submitted to FM89, Conference on the Use of Formal Methods in Systems Design, Halifax, July 1989.

[15] W. J. Cullyer. Implementing Safety Critical Systems: The VIPER Microprocessor. In G. Birtwistle and P. A. Subrahmanyam, editors, *VLSI Specification, Verification and Synthesis*, pages 1–26, Norwell, Massachusetts, 1988. Kluwer.

[16] I. S. Dhingra. *Formal Validation of an Integrated Circuit Design Style*. PhD thesis, University of Cambridge Computer Laboratory, 1988.

[17] A. J. Field and P. G. Harrison. *Functional Programming*. Addison–Wesley, New York, 1988.

[18] M. J. C. Gordon. *The Denotational Description of Programming Languages*. Springer Verlag, London, 1979.

[19] M. J. C. Gordon. LCF-LSM: A System for Specifying and Verifying Hardware. Technical Report 41, Computing Laboratory, University of Cambridge, 1983.

[20] M. J. C. Gordon. Proving a Computer Correct using the LCF-LSM Hardware Description Language. Technical Report 42, Computing Laboratory, University of Cambridge, September 1983.

[21] M. J. C. Gordon. Why higher-order logic is a good formalism for specifying and verifying hardware. In G. Milne and P.A. Subrahmanyam, editors, *Formal Aspects of VLSI Design*, pages 153–177, Amsterdam, 1986. North-Holland.

[22] M. J. C. Gordon. HOL: A proof generating system for higher-order logic. In G. Birtwistle and P. A. Subrahmanyam, editors, *VLSI Specification, Verification and Synthesis*, pages 73–128, Norwell, Massachusetts, 1988. Kluwer.

[23] M. J. C. Gordon. *Programming language theory and its implementation*. Prentice Hall, London, 1988.

[24] B. Graham. Dealing with the Choice Operator in HOL88. Research Report 90/382/06, Computer Science Department, University of Calgary, 1990. Available in the *contrib* directory of the HOL system.

[25] B. Graham. From Synchronous Microprocessor to Asynchronous Circuits. Unpublished research report for the Computer Science Laboratory, University of Cambridge, July 1991.

[26] B. Graham, S. Williams, and G. Stone. Operating Specification for the SECD Chip. Research Report 89/353/15, Computer Science Department, University of Calgary, 1989.

[27] B. T. Graham. SECD: The Design and Verification of a Functional Microprocessor. Master's thesis, University of Calgary, June 1990. Research Report No. 90/395/19, Department of Computer Science, University of Calgary.

[28] P. Henderson. *Functional Programming; Applications and Implementation.* Prentice Hall, London, 1980.

[29] P. Henderson, G. A. Jones, and S. B. Jones. The Lispkit Manual, volume 1. Technical Monograph, PRG-32(1), Oxford University Computing Laboratory, 1983.

[30] P. Henderson, G. A. Jones, and S. B. Jones. The Lispkit Manual, volume 2. Technical Monograph, PRG-32(2), Oxford University Computing Laboratory, 1983.

[31] M. J. Hermann, G. Birtwistle, B. Graham, and T. Simpson. The Architecture of Henderson's SECD Machine. Research Report 89/340/02, Computer Science Department, University of Calgary, 1989.

[32] W. A. Hunt. FM8501: a Verified Microprocessor. Technical Report 47, Computer Science Department, University of Austin at Texas, Austin, Texas, 1985.

[33] J. Joyce. The SECD Machine, A Study in Advanced Architectures. Unpublished report of cpsc 603 course work at the University of Calgary.

[34] J. Joyce. Formal Verification and Implementation of a Microprocessor. In G. Birtwistle and P. A. Subrahmanyam, editors, *VLSI Specification, Verification and Synthesis*, pages 129–157, Norwell, Massachusetts, 1988. Kluwer.

[35] J. Joyce. Case Study: Microprocessor Systems, 1989. In The HOL System: Tutorial, pages 115–232.

[36] J. Joyce. Multi-Level Verification of a Simple Microprocessor. Progress Report, December 1989.

[37] J. Joyce. A Verified Compiler for a Verified Microprocessor. Technical Report 167, University of Cambridge Computer Laboratory, 1989.

[38] P. J. Landin. The mechanical evaluation of expressions. *The Computer Journal*, 6(4):308–320, 1964.

[39] M. Leeser and G. Brown, editors. *Hardware Specification, Verification, and Synthesis: Mathematical Aspects*. Springer-Verlag, New York, 1989. Proceedings of the Mathematical Sciences Institute Workshop, Cornell University, Ithaca, N.Y., July, 1989.

[40] Ian A. Mason. *The Semantics of Destructive LISP*. Center for the Study of Language and Information, 1986.

[41] Ian A. Mason. Verification of Programs that Destructively Manipulate Data. *Science of Computer Programming*, 10:177–210, 1988.

[42] Carver Mead and Lynn Conway. *Introduction to VLSI Systems*. Addison-Wesley Publishing Company, 1980.

[43] T. F. Melham. Abstraction Mechanisms for Hardware Verification. In G. Birtwistle and P. A. Subrahmanyam, editors, *VLSI Specification, Verification and Synthesis*, pages 267–291, Norwell, Massachusetts, 1988. Kluwer.

[44] J. Strother Moore. PITON: A Verified Assembly Level Language. Technical Report CLI-22, Computational Logic Inc, Austin, Texas, 1988.

[45] P. Narendran and J. Stillman. Formal Verification of the Sobel Image Processing Chip. In G. Birtwistle and P. A. Subrahmanyam, editors, *Current Trends in Hardware Verification and Automated Theorem Proving*, pages 92–127, New York, 1989. Springer Verlag.

[46] Peter G. Neumann. Illustrative Risks to the Public in the Use of Computer Systems and Related Technology. Position paper submitted to FM89, Conference on the Use of Formal Methods in Systems Design, Halifax, July 1989.

[47] G. D. Plotkin. Call-by-name, call-by-value, and the lambda calculus. *Theoretical Computer Science*, 1(1):125–159, 1975.

[48] S.M. Rubin. *Computer Aids to VLSI Design*. Addison-Wesley, Reading, MA, 1987.

[49] R. C. Sekar and M. K. Srivas. Formal Verification of a Microprocessor Using Equational Techniques. In G. Birtwistle and P. A. Subrahmanyam, editors, *VLSI Specification, Verification and Synthesis*, pages 171–217, New York, 1989. Springer Verlag.

[50] T. Simpson, G. Birtwistle, B. Graham, and M. J. Hermann. A Compiler for Lispkit Targetted at Henderson's SECD machine. Research Report 89/339/01, Computer Science Department, University of Calgary, 1989.

[51] T. Simpson, B. Graham, and G. Birtwistle. From LispKit to SECD Chip: Some Steps on the way to a Verified System. Proceedings of the Third Banff Verification Workshop, 1989, 1989.

[52] K. Slind, G. Birtwistle, M. Hermann, and T. Simpson. From Specification to Layout: Transforming HOL Specifications into Gate Array Net-Lists. In *Proceedings of Canadian Conference on Electrical and Computer Engineering*, 1989.

[53] J. Staunstrup, editor. *Formal Methods for VLSI Design*. North Holland, 1990. IFIP WG 10.5 Summer School (Lyngby, 1990).

[54] M. Williams. SECD Controller Board Implementation. Technical Report #89/359/21, Computer Science Department, University of Calgary, Calgary, 1989.

[55] W. D. Young. A Mechanically Verified Code Generator. Technical Report CLI-37, Computational Logic Inc, Austin, Texas, 1988.

Index

⊢, 5